Clear
Concise
Complete

商務實戰
英語書信實例

Practical Model Business Correspondence

U0119310

Practical Model Business Correspondence

Clear · Concise · Complete

Contents

Practical Model Business Correspondence

Clear · Concise · Complete

Contents

Practical Model Business Correspondence

Clear · Concise · Complete

Contents

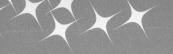

作者的話

英文的重要性：

　　本人以理工學歷背景投入職場後才驚覺英文的重要性竟遠大於自身的專業技能。單憑高工學歷者便可從數十、百位工程師中脫穎而出，當上外商公司的技術部經理、品管主任的人不在少數，原因在於這些人都能以流利的英文和外國客戶溝通。

出書的目的：

　　希望藉由這本英文工具書的出版及推廣，提昇讀者個人在職場、公司在產業界，甚至台灣整體產業在國際上的競爭力及專業形象。基本上，語文的使用是源自於模仿而非自創。當要寫一篇自己從未讀過或寫過的英文文章是很難下筆的，就算勉強完成，充其量只是充滿中文語法、不太通順的字詞組合，且別人不一定看得懂的中文式英文（Chinglish）。有鑑於此，本書特地收錄職場上各類實用的商務主題，格式涵蓋一般書信、傳真、公告、備忘錄、合約、電子郵件等經典實務範例，無論你是要發封簡短的祝賀信、下訂單、報價、撰寫聘雇函，或擬定正式的代理合約，本書都將是你最寶貴的參考資源。

本書特色：

1. 大部分的範例是由在業界擁有十幾年經驗的資深美國工程師及專業經理人所撰寫，並加以編輯、潤飾，加入更多實用的英文句型與片語，讀者將可學到職場實務、通用的美式英文，跟坊間所買到用制式、學院派英文所編纂而成的商業書信文章截然不同。

2. 市面上此類工具書的場景大多以美國為主，然而本書內容是以台灣出發，包含與美國

總公司、外國客戶等日常聯繫、議價、安排參訪等內容，並收錄許多和台灣周遭環境相關的英文：風俗民情、企業經營環境、勞基法、海關規定、簽證、專利申請、銀行作業等。

3. 文章涵蓋不同產業別，如傳統產業、高科技電子業等。除了書信寫作技巧外，同時教授許多實用的商業用語，如 freight collect、L / C at sight、pilot run、 zero-hour quality 等，閱讀本書就如同取得某種程度的專業工作資歷。

後記：

　　本書歷經三年的籌劃準備，今得以付梓上市，非常感謝海內外許多商業夥伴的傾囊相授，更感謝 LiveABC 編輯團隊秉持精益求精、棄而不捨的工作態度使本書更具參考價值、實用性與豐富度。內容如有不盡完備之處尚請讀者來函，或於部落格留言指正。不勝感激！

電子信箱：
jm11290@hotmail.com

部落格：
http://tw.myblog.yahoo.com/win-english

作者簡介：莊錫宗

歷任：

1984 ~ 1986
荷商生產工程師

1986 ~ 1989
和來自 **14** 個先進國家的工程師在印尼參與總價 **10** 億美金的冷軋鋼廠試俥計畫

1989 ~ 1995
美商公司品管經理
技術部經理
台灣區聯絡辦事處經理

1995 ~
開設貿易公司
從事商業英文教學

Chapter One

Office

Unit One
Human Resources
人力資源

本單元是有關人事方面的文章，內容包含推薦信函、自我簡介、聘雇確認書、未錄取通知信函、薪資等。

 Learning Goals

1. 了解推薦信函的寫法

2. 學習撰寫自我簡介，及其與履歷表或自傳不同之處

3. 認識與人事聘雇、薪資等相關之常用詞彙

Before We Start 實戰商務寫作句型

Topic 1: Recommending Someone 推薦人才

◆ He's proven / shown himself to be [a valuable employee / a reliable worker / a leader / an asset to this company . . .].
他證明自己是位 [有價值的職員／可信賴的員工／領袖／這間公司的寶貴人才……]。

◆ His [computer / communication / management . . .] skills are [superior to others / exceptional / first-rate / top-notch / impressive . . .].
他的 [電腦／溝通／管理……] 技巧 [優於他人／一流／令人印象深刻……]。

◆ From my experience / In my opinion / From my point of view, S. + V. . . .
依我（的經驗）看來……

◆ I highly recommend (someone) for the position of . . .
我極力推薦（某人）擔任……的職務。

Topic 2: Informing Someone of Something 通知消息

◆ I'm writing to tell / inform you that . . .
This is to make you aware / inform you that . . .
I want to fill you in on . . .
I want to let you know that . . .
Please be advised that . . .
我（寫信）要告訴你……

A. Recommendation Letter 推薦信函

本篇文章是 Jack Huang 在加入 Maples Taiwan 前，於印尼分公司工作快結束時，部門的美籍顧問幫他寫的推薦信函。

To Whom It May Concern:

I have had the pleasure of working closely with Mr. Jack Huang for the past two years at the CRMJU plant site in Merak, Indonesia. During this time, Mr. Huang has proven himself to be an **extremely**[1] valuable employee. His technical skills and knowledge of annealing* **are superior to**[2] those of most people, and his work attitude has been **exceptional**.[3]

Mr. Huang has also shown **remarkable**[4] **adaptability**[5] to the multicultural environment of Merak. He speaks both English and Indonesian fluently. Finally, Mr. Haung has displayed outstanding leadership skills, both with his **crew**[6] and other **expatriates**[7] alike.

I highly recommend Mr. Huang for a position of equal or greater responsibility. I am confident that he will be an **asset**[8] to any organization.

Yours faithfully,
T.H. Moorse
T.H. Moorse
UEC — Consultant
Continuous Annealing Department
P.T. CRMJU — Merak, Indonesia

中譯

七月二十號

敬啟者：

我非常榮幸過去兩年來在印尼孔雀港的爪哇冷軋鋼廠（**CRMJU**），與黃傑克先生密切地共事。這段期間，黃先生的表現證明了他是一位非常有價值的員工。他的專業技能與在冷軋退火方面的知識凌駕大多數人，他的工作態度也非常好。

黃先生在孔雀港當地多元文化的環境中也展現出他卓越的適應力，他的英文和印尼文都講得很流利。最後，對於跟他一起工作的組員與其他外籍人士而言，他也展現了卓越的領導才能。

我極力推薦黃先生擔任其他同等或更高階的職位，我確信黃先生對任何組織而言都會是相當寶貴的人才。

謹啟
T.H. 摩斯
UEC 顧問
連續退火部門
爪哇冷軋鋼廠有限公司 — 印尼孔雀港

14

Vocabulary and Phrases

1. **extremely** [ɪkˋstrimlɪ] *adv.* 非常地
   I had just finished working a 16-hour day and was extremely tired.
   我剛結束十六個小時的工作，非常地疲倦。

2. **be superior to** 比……好的；勝過……的
   He thinks French wine is superior to South African wine.
   他覺得法國的酒勝過南非的酒。

3. **exceptional** [ɪkˋsɛpʃənḷ] *adj.* 優秀的；卓越的
   The news report was exceptional.
   那篇新聞報導寫得很好。

4. **remarkable** [rɪˋmɑrkəbḷ] *adj.* 驚人的；非凡的
   Last year's sales growth was remarkable.
   去年的業績成長很驚人。

5. **adaptability** [ə͵dæptəˋbɪlətɪ] *n.* 適應性
   You're smart, but we're worried about your adaptability.
   你很聰明，但我們擔心你的適應力。

6. **crew** [kru] *n.* 一組工作人員；同事
   She has six people on her crew.
   她的團隊有六個人。

7. **expatriate** [ɛksˋpetrɪ͵et] *n.* 移居國外者
   I've heard that 80 percent of Dubai's workers are expatriates.
   我聽說杜拜百分之八十的員工都是外籍人士。

8. **asset** [ˋæ͵sɛt] *n.* 寶貴的人才；資產
   It's too bad Wayne quit. He was a big asset to this company.
   韋恩辭職實在很可惜，他是這間公司非常寶貴的人才。

Biz Focus

★ **annealing** 鐵、鋼或玻璃等製程中的退火（用於消除殘留的內部應力）

Language Corner

How to Start a Recommendation Letter

I have had the pleasure / honor of working closely with . . .

You Might also Say . . .

- Over the past few years, I have come to appreciate . . .
- I am writing on behalf of . . .
- As a colleague / the immediate supervisor of . . .

B. Background Sketch 自我簡介

新任總經理想要認識公司的高階主管，於是透過秘書 Betty 請同仁們繳交自我簡介。

Oct. 2

TO: BETTY

Here is the background sketch you requested: Mr. Raymond (Richard) Barlow, Director of Foreign Operations.

Mr. Barlow studied **electronics**[1] at the Naval Academy in Annapolis and served as an air crewman in an antisubmarine warfare patrol squadron until 1977. Mr. Barlow then went on to study engineering at Berkeley University before spending seven years with Lette Company as a test engineer, six years with the Twin Brothers **division**[2] of Central Mills as the reliability engineering* manager, and a year with Lico Enterprises as Corporate Manager of Safety and Reliability Engineering. In 1996, he joined Maples Corporation as Director of Product Integrity. Two years later, Mr. Barlow **assumed the role of**[3] Director of Far East Operations, and in 2000 became Director of Foreign Operations. His background includes **extensive**[4] QC* / QA* and **manufacturing**[5] engineering experience. He is a member of the American Society for Quality and the American Society of Test Engineers.

中譯

十月二號

收件人：貝蒂

這是妳所要求的個人背景簡介：雷蒙（理查）·巴洛先生 ─ 海外營運部總監。

巴洛先生在美國安那波理斯的海軍學院研讀電子學，並擔任反潛戰鬥巡邏艦隊的空勤人員直至 1977 年，之後到柏克萊大學唸工程。巴洛先生在 Lette 公司任職七年，擔任測試工程師；在 Central Mills 的分支機構 Twin Brothers 當了六年的可靠度工程經理；之後到 Lico 企業當了一年的安全與可靠度工程經理。1996 年始加入 Maples 公司擔任產品品質總監。兩年後巴洛先生擔任遠東區營運部總監，2000 年成為海外部總監。他擁有豐富的品管／品保與製造工程的經歷，他同時也是美國品管學會與美國測試工程師學會的會員。

Vocabulary and Phrases

1. **electronics** [ɪ͵lɛk`trɑnɪks] *n.* 電子學

   Jeremy spends most of his time studying electronics.
   傑若米大部分的時間都花在研讀電子學。

2. **division** [də`vɪʒən] *n.* （機關、公司等的）部門；處；課

   I've never heard of that team before. What division do they work in?
   我以前從沒聽過那個團隊，他們在哪個部門工作？

3. **assume the role of** 擔任⋯⋯的職務

   Mr. Peters assumed the role of vice president last summer.
   皮德斯先生去年夏天接下副總裁的職務。

4. **extensive** [ɪk`stɛnsɪv] *adj.* 廣泛的；大量的

   We're looking for someone with extensive experience in marketing.
   我們正在找有豐富行銷經驗的人。

5. **manufacturing** [͵mænjə`fæktʃərɪŋ] *n.* 製造業

   The manufacturing sector has really taken off in India.
   印度的製造業開始蓬勃發展。

Biz Focus

★ **reliability engineering** 可靠度工程

「可靠度」是指產品的品質能滿足顧客使用時的需求。產品設計時經由預防與驗證，將「可靠度」設計於產品品質中，使產品具有適用、穩定與精準等特性。而「可靠度工程」就是利用統計學、機率、韋伯分佈等，研究產品在特定時間內達到某種運作要求的機率，常見的指標值有「失效率」和「平均壽命」。

★ **QC** 品管（= quality control）

不管是進料檢驗（incoming quality control）、製程檢驗（in-process quality control）或出貨檢驗（outgoing quality control），都是用來檢測不同階段所製造出來的品質，所採取的一種定期而隨機的品質管制措施。

★ **QA** 品保（= quality assurance）

指廣義的品質管理工作，從對外包含協力廠商及客戶，到對內包含新產品的設計研發、生產製造、出貨和售後服務等，均需建立全面性的品管體系。品保是事先預防的工作，期望能「第一次就做對」，而品管則是事後補救的檢驗工作。

Insight

Company / Enterprise / Corporation 用法大不同

	定義	用法
Company	公司	多用於法律資料中，指「社團、法人、公司、企業」等。
Enterprise	（通常指從事商業投資的）企業	當公司名稱時，字首通常會大寫並加上 s。
Corporation	（美）股份（有限）公司	「有限」是指公司股東償還債務的責任（liability）是有限的，萬一公司負債，最多賠掉資本額，股東不必再自掏腰包賠錢。

COMPANY RESOURCES

在不同國家，「有限公司」的表達法亦有所不同：

英國	Co., Ltd.（Company Limited）
美國	Corporation 或 Inc.（Incorporated）
澳洲	Pty. Ltd.（Proprietary Limited）
德國	GmbH
印尼	P.T.

註：台灣以英式寫法為主。

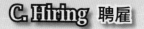

聘雇函應包含工作職稱、到職日期、薪資福利、試用期等資訊。信件最後也會附上簽名欄，請應試者回覆是否接受。

I. Employment Offer Letter 聘雇函

MAT / MAPLES CORPORATION

No. 305, Sec. 2, Chien Kuo N. Rd., Taichung City, Taiwan, ROC

e-mail:personnel@maples.com   Tel: 886-4-2356-0877   Fax: 886-4-2356-0800

August 3rd

Jack Huang
6F, No. 32, Sec. 3, Bade Rd.
Songshan, Taipei 105, Taiwan

Dear Mr. Huang,

I am writing to tell you I am pleased to confirm our offer of a **permanent**[1] position with Maples Corporation as a quality control engineer starting on September 2nd.

I am **enclosing**[2] the **terms**[3] we discussed during the interview:

- Start Date: September 2nd
- Salary:

Your starting salary will be NT$35,000 per month, to be reviewed after a three-month **probationary period**.[4] If you are successful in completing this phase of employment, your salary will be increased to NT$38,000 a month.

- Benefits:

We offer health, labor, and dental insurance, seven days of annual leave, and **contributions**[5] to a retirement fund. Employees **are** also **entitled to**[6] a yearly bonus of up to three months.

Your **compensation package**[7] will thereafter be **reassessed**[8] on an **annual**[9] basis.

We're all hoping you join the team and we look forward to working with you.

Sincerely,

Edward Smith
Edward Smith
Personnel Manager

- -

I would like you to peruse the contents of this employment offer letter. To accept, please sign and date the letter below and fax it back to MAT before August 17th.

I, Jack Huang, agree to accept the position offered by Maples Corporation and the terms indicated above.

_____   _____
(Employee's Signature)   (Date)

Note: If you wish to decline this job offer, please reply with a brief letter.

中譯

八月三號

黃傑克
105 台灣台北市松山區八德路三段三十二號六樓

親愛的黃先生：

此封信是要告訴您我很榮幸向您確認，我們願意提供您在 Maples 公司擔任品管工程師的全職職務，從九月二號起生效。

謹附上我們面試時所討論的條件：

● 開始上班日期：九月二號

● 薪水：
您的起薪為每月新台幣三萬五千元，三個月試用期過後會做評估，如果您成功地完成這個階段的聘雇，您的月薪會增加為三萬八千元。

● 福利：
我們提供勞保健保和牙醫保險、七天年假及退休金提撥。員工可享多達三個月的年終獎金。

之後每年會再評估您的薪水與福利。

我們全都希望您能加入我們的團隊，並期待能與您共事。

祝 商安
艾德華・史密斯
人事部經理

我要請您仔細閱讀這封聘雇函的內容。如果您願意接受，請在下面簽名、標上日期，並於八月十七號前將信件傳真到 MAT。

我（黃傑克）同意接受 Maples 公司提供的職務以及上述的條件。

_____   _____
（員工簽名）   （日期）

註：如果您要拒絕這個工作機會，請簡短回覆。

Language Corner

How to Start an Offer Letter:

I am writing to tell you I am pleased to confirm our offer of . . .

You Might also Say . . .

- We would like to offer you the position of . . . starting on / effective . . .
- (Company) is pleased to offer you a job as . . .
- In conclusion to the interview we had on (date), we're providing a permanent / temporary / seasonal position of . . .

Vocabulary and Phrases

1. **permanent** [`pɝmənənt] *adj.* 永久的；固定性的
   Edward has been a permanent fixture in this company for years.
   艾德華是在這間公司工作好幾年的老員工。

2. **enclose** [ɪn`kloz] *v.* 附上；裝入
   I have enclosed one copy of the contract.
   我附上一份合約的副本。

3. **terms** [tɝmz] *n.* （合約等的）條件；條款
   The terms of this agreement are not favorable to my company.
   這份協議的條件不利於我的公司。

4. **probationary period** 試用期
   If you're hired, your probationary period will last three months.
   如果你被雇用，你會有三個月的試用期。

5. **contribution** [ˌkɑntrə`bjuʃən] *n.* 分擔額
   The company makes a monthly contribution to the pension plan on your behalf.
   公司每月替你提撥退休金。

6. **be entitled to** 有權；可享有
   Only full-time employees are entitled to health and labor insurance.
   只有全職員工才能享有勞健保。

7. **compensation package** 薪資組合（含薪水及福利等）
   I believe our compensation package is more than fair.
   我認為我們的薪資組合真的很公平。

8. **reassess** [ˌriə`sɛs] *v.* 對……再評估
   Mr. Jenkins wants you to find out what is going on and reassess the situation.
   詹金先生要你了解狀況，並重新評估情勢。

9. **annual** [`ænjʊəl] *adj.* 每年的；一年一次的
   I file my taxes on an annual basis.
   我每年都有報稅。

在決定錄取人選後，公司應儘快通知未錄取者，謝謝他們投遞履歷並前來面試。

II. Rejection Letter 未錄取通知

#1 應徵者在第一階段的資料審核就被刷下。

To: Teresa Foster
From: Harold Sintich
Date: May 10
Subject: Re: Application

Dear Ms. Foster,

Thank you for your response to our recent advertisement offering a management trainee position. Your **credentials**[1] are **impeccable**,[2] and I am sure you would be a **credit**[3] to any company with which you are associated.

Although our current **vacancy**[4] has been filled by another candidate, we will **keep** your application **on file**[5] in case we have other opening in the future.

Thanks again for your application, and good luck with your job search.

Harold Sintich

#2 應徵者雖然得以參加面試，但仍未被錄取。

To: Rob Gardenhire
From: Sophia Wu
Date: August 21st
Subject: Keep in Touch

Dear Mr. Gardenhire,

I am writing at this time to inform you that we at Maples Corporation have decided to go with another applicant. From your interview, we have concluded you are a **first-rate**[6] candidate. If it is OK with you, we would like to keep your CV **for the time being**.[7]

I do not mind telling you this: you were one of the most **promising**[8] candidates we had during this round

中譯

#1
收件人：德蕾莎·佛斯特
寄件人：哈洛德·辛堤志
日期：五月十號
主旨：回覆：應徵

親愛的佛斯特小姐：

謝謝您回覆我們最近找儲備幹部的徵才廣告，您的資歷非常完美，我相信您必定會為您所任職的公司增光。

雖然我們已經找到另一位人選來擔任該職位，不過我們仍會將您的申請資料留存備查，以備我們將來還有其他職缺。

再次感謝您的應徵，並祝您找工作順利。

哈洛德·辛堤志

#2
收件人：羅伯·葛登艾
寄件人：吳蘇菲亞
日期：八月二十一號
主旨：保持聯繫

親愛的葛登艾先生：

我此時寫信給您是要通知您 Maples 公司已經決定錄取另一位應徵者。從與您的面談中，我們得知您是一流的人選，如果您同意的話，我們希望能保存您的履歷表。

不瞞您說：「您是我們這一輪面試中最有前途的人選之一。」我想要

of interviews. I want to personally thank you for taking the time to come in for an interview.

We wish you all the best in your job hunt.

Sophia Wu
Accounting Manager

親自謝謝您撥空來參加面試。

我們祝您找工作順利。

吳蘇菲亞
會計經理

Vocabulary and Phrases

1. **credentials** [krɪ`dɛnʃəlz] *n.* 資歷；憑證
   This applicant's credentials are a bit weak.
   這位應徵者的資歷有點不足。

2. **impeccable** [ɪm`pɛkəbḷ] *adj.* 無懈可擊的；無缺點的
   Besides having impeccable references, Bobby also has a lot of experience in this field.
   除了有很好的推薦信，巴比在這個領域的經驗也很豐富。

3. **credit** [`krɛdɪt] *n.* 增光的人（事）；光榮
   Josh is a credit to his parents.
   賈許的父母以他為榮。

4. **vacancy** [`vekənsɪ] *n.* 職缺
   Does your company have vacancies right now?
   你們公司現在有職缺嗎？

5. **keep sth on file** 將……建檔、留存備查
   We'll keep your resume on file for six months.
   我們會將你的履歷表保存六個月。

6. **first-rate** [`fɜst`ret] *adj.* （口語）極好的；優秀的
   As Karen is a first-rate accountant, we should give her a raise.
   由於凱倫是位很棒的會計，我們應該給她加薪。

7. **for the time being** 目前；暫時
   We need you to work in outside sales for the time being.
   我們需要你暫時作外務推銷的工作。

8. **promising** [`pramɪsɪŋ] *adj.* 有前途的；大有可為的
   If Sean can stay focused, I think he'll be a very promising young executive.
   如果尚恩可以集中精力，我認為他會是位很有前途的年輕主管。

Language Corner

How to Write a Good Rejection Letter

1. 感謝對方投遞履歷、參加面試
   - Thank you for responding to our advertisement for (position).
   - We appreciate your applying for the position of . . . at our company.

2. 說明未錄取的原因
   - This position requires . . .
   - We've decided to go with another applicant at this time.

3. 善用附屬連接詞，如 although、while 等
   - Although we've decided upon another candidate, we will still . . .
   - While we were impressed with your credentials, we have determined that . . .
   - Although you have made some concessions, we still cannot . . .

4. 被動語態的語氣較為婉轉
   - Unfortunately, an offer has been made to another candidate.
   - Another individual has already been brought on board.
   - The position has already been filled internally.

5. 禮貌性作結
   - Thank you again for your interest in our company. We wish you success for your future career.
   - Your credentials are impeccable. I am sure you will not have trouble securing a position in another company.

21

D. Payroll 薪資

Maples 公司在大陸請了一位品管工程師負責當地的驗貨工作，但由於當地法令的限制，無法直接支薪給該名員工，於是商請當地的代工工廠（Global OEM）代墊薪資。

#1 告知財務部門。

TO: Eric Walker
FROM: Daniel Harris
DATE: 6/22
SUBJECT: New Employee

Eric,

We've hired an engineer in mainland China to **inspect**[1] production at Global OEM, and we can't pay him directly because of China's regulations. We've made arrangements with GO to put him on their payroll and **invoice**[2] us for the amount.

The purpose of this letter is to inform you that invoices you receive from GO for our guy are **legitimate**[3] in the amount of **approximately**[4] US$250 per month, which may **fluctuate**[5] due to changes in the **exchange rate**.[6]

Sincerely,
Daniel

#2 與 OEM 工廠溝通相關發票問題。

TO: Patty Lo
FROM: Daniel Harris
DATE: 9/23
SUBJECT: Payroll

Dear Ms. Lo,

Per our phone conversation, Harry, our engineer in your China factory, has told me that starting September 1st, he will no longer eat at the GO cafeteria. Please **adjust**[7] your invoicing to Maples Corporation to **reflect**[8] this change, which is US$35 less each month.

Thanks,
Daniel

中譯 #1

收件人：艾瑞克·渥克
寄件人：丹尼爾·哈利斯
日期：六月二十二號
主旨：新進員工

艾瑞克：

我們在大陸請了一位工程師來負責 Global OEM 的驗貨工作，但由於當地的法令限制，我們無法直接支付他薪水，因此我們和 GO 協調，請他們付他薪水，再開發票給我們。

這封信的目的是要告知你，你所收到由 GO 開立關於該員工的發票是合法的，每月大約有兩百五十元美金，金額會隨匯率而有所變動。

祝 商安
丹尼爾

中譯 #2

收件人：羅佩蒂
寄件人：丹尼爾·哈利斯
日期：九月二十三號
主旨：薪資

親愛的羅女士：

如我們在電話上談過的，我們派駐在你們大陸工廠的工程師哈利斯告訴我，從九月一號開始他就不會再去 GO 的自助餐廳用餐了，請調整你們開給 Maples 公司的發票以反映此項改變，每月將減少三十五元美金。

謝謝
丹尼爾

Vocabulary and Phrases

1. **inspect** [ɪnˈspɛkt] *v.* 檢查

   Upon inspecting the shipment, we found it to be short on several items.
   在檢查那批貨時，我們發現缺少好幾個品項。

2. **invoice** [ˈɪn͵vɔɪs] *v.* 開發票給……
   (*n.* 發票；發貨單)

   Look! I've been invoiced on this product twice.
   你看！這個產品我被重複開了兩次發票。

3. **legitimate** [lɪˈdʒɪtəmɪt] *adj.* 合法的；正當的

   Is all of this paperwork legitimate, Charles?
   查爾斯，這所有的文件都合法嗎？

4. **approximately** [əˈprɑksəmɪtlɪ] *adv.* 大概；大約

   The plane will be landing in Frankfurt at approximately 7:25 this evening.
   那班飛機大約今晚七點二十五分會降落在法蘭克福。

5. **fluctuate** [ˈflʌktʃʊ͵et] *v.* 變動

   The US dollar has been fluctuating so much recently that we don't know how to set the price.
   美金最近的變動很大，以致於我們都不知道該如何定價。

6. **exchange rate** 匯率

   I want to buy 50,000 baht. What's the exchange rate?
   我想買五萬元的泰銖，匯率是多少？

7. **adjust** [əˈdʒʌst] *v.* 調整

   Could you adjust the terms on this contract to include this item?
   你可以調整這份合約的條件，以包含這項條款嗎？

8. **reflect** [rɪˈflɛkt] *v.* 反映；表現

   This invoice still does not reflect all of the charges.
   這張發票還是沒有顯現所有的費用。

Language Corner

1. Talking about Currency Changes

The amount may fluctuate due to changes in the exchange rate.

You Might also Say . . .

- The amount is still vulnerable to currency fluctuations.
- Fluctuating currencies may cause this amount to go up or down at any time.

2. How to Follow Up

Per our phone conversation, . . .

You Might also Say . . .

- To run over / reconfirm what we talked about on the phone, . . .
- I need to double-check on what we discussed on the phone.
- I'm following up on our phone call. Did you mean . . . ?
- When you said . . . on the phone this morning, what did you mean?

Unit Two
Administration 辦公室管理

當公司出現人事異動時，為求日常運作不受影響，必需重新分配工作任務，同時宣達公司內部相關規定及員工應遵守事項。

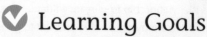

✅ Learning Goals

1. 了解在有人離職及人力資源有限的情況下，如何有效率地分配工作

2. 學習如何制定並公告員工守則

Before We Start 實戰商務寫作句型

Topic 1: Assigning Someone a Task 分配工作

* (Someone) will be responsible for / be in charge of / be taking care of . . .
  （某人）將負責……

* (Someone) will assume the role of (position).
  （某人）將接……的職務。

* (Someone) has assumed a number of responsibilities in . . .
  （某人）擔任……方面的一些職務。

* (Someone) will cover the [switchboard / reception desk . . .] for . . .
  （某人）將代理 [總機 / 接待櫃檯……] 的職務。

Topic 2: Requiring a Supervisor's Approval 報請主管核准

* [Break times / professional development programs / outings . . .] will be scheduled / organized / arranged at the discretion of each employee's supervisor.
  [休息時間 / 專業進修課程 / 員工旅遊……] 將由每個員工的主管來安排時間 / 規劃 / 籌備。

* [Overtime / reimbursements / bonuses . . .] will only be paid once your supervisor has signed for / approved it (them).
  需經主管簽核 / 批准，使得支領 [加班費 / 補償金 / 獎金……]。

* Requests for [time off / stationery / comp time . . .] should be made to / through / via your supervisor.
  需向 / 經主管提出 [休假 / 文具 / 加班補休……] 的申請。

A. Reallocation of Tasks 重新分配工作

Maples 公司的營運部副總 Ted 發公告，說明營運部經理 Eric 離職後的職務分配。

○ ○ ○

To: All Employees
From: Ted Parker
Date: 4/26
Subject: Promotions

With the **departure**[1] of Eric Woods, Edith Ruby has been promoted to Manager of Operations. She will be assisting me in **coordinating**[2] various aspects of the production and shipment of orders, communicating with customers and manufacturers, and ensuring that the production runs as smoothly as possible. In addition, she will be responsible for the day-to-day operations of the office.

Shelly Hoffman has assumed a number of additional responsibilities in our purchasing and financial areas, specifically **with regard to**[3] purchase order control and L / C **administration**.[4]

With these added responsibilities, neither Edith nor Shelly will be able to commit themselves to covering receptionist breaks.

Therefore beginning immediately, coverage of the switchboard will be as follows:

1. The receptionist will handle the switchboard from 9:00 a.m. to 5:30 p.m.
2. During his / her lunch breaks, Betty and Iris will take turns answering the phones.
3. Additional coverage of the switchboard may be necessary at times when we send the receptionist on errands. Ideally, we will know of these situations **in advance**[5] and will notify Betty and Iris that their help is needed. In order to make a smooth **transition**,[6] we are counting on the cooperation of all.

Concerning the copying of the faxes:

For the time being, all incoming faxes will be copied by either Betty or Iris first thing in the morning. It is **imperative**[7] that all parties involved read their faxes at the fax table until all copies are made. This will ensure that all incoming faxes are **accounted for**.[8]

Thanks,
Ted

中譯　收件人：全體員工
寄件人：泰德‧帕克
日期：四月二十六號
主旨：晉升

　　隨著艾瑞克‧伍茲的離職，伊迪絲‧茹比被升任為營運部經理，她將協助我協調各種生產和訂單出貨事宜、與顧客及製造商溝通，並確保生產得以順利進行。此外，她將會負責辦公室的日常營運。
　　雪莉‧霍夫曼兼負一些採購與財務方面的職責，尤其是有關採購單控管與信用狀管理事宜。

由於這些增加的工作量,伊迪絲和雪莉都將無法在總機休息的空檔幫忙接聽電話。

因此總機的工作從現在起將如下分配:

1. 接待員負責總機時間為早上九點到下午五點半。
2. 總機午休時,貝蒂和艾瑞絲會輪流接聽電話。
3. 當我們派總機出去辦事時,可能需要額外支援。理論上,我們可以事先得知這些狀況,並且通知貝蒂和艾瑞絲來幫忙。為了能順利度過過度期,我們將仰賴大家的幫忙。

有關影印傳真:

目前每天一大早貝蒂或艾瑞絲就會影印所有傳來的傳真,在所有的傳真影印完畢之前,所有相關人等都只能在傳真機旁看傳真,這點十分重要,這樣才能確保所有傳來的傳真都有被記錄到 。

謝謝
泰德

Vocabulary and Phrases

1. **departure** [dɪ`pɑrtʃɚ] *n.* 離開;出發

   Vince usually makes an early departure from the office on Friday afternoons.
   文斯星期五下午通常都會提早離開辦公室。

2. **coordinate** [ko`ɔrdn̩et] *v.* 協調

   Roseanne will be coordinating the conference.
   蘿珊將協調這個會議的相關事宜。

3. **with regard to** 關於

   With regard to what the manager has said about recycling, I think it's a good idea.
   有關經理提到的回收事宜,我認為是個好主意。

4. **administration** [ədˌmɪnə`streʃən] *n.* 管理

   I'm really pleased by how well the office administration has been running.
   我很滿意辦公室管理得很好。

5. **in advance** 事先;預先

   I'll let you know two weeks in advance.
   我會在兩星期前讓你知道。

6. **transition** [træn`zɪʃən] *n.* 過度期;轉變

   The company is in the middle of a transition. That's why there have been changes.
   這家公司正處於過度期,所以會有些改變。

7. **imperative** [ɪm`pɛrətɪv] *adj.* 極重要的;必要的

   It's imperative that we have that shipment by next Thursday.
   我們必需在下星期四以前收到那批貨。

8. **account for** 記錄;知道……的下落

   Every dollar spent needs to be accounted for.
   所花的每一塊錢都需要記錄下來。

Language Corner

1. How to Move Along

| With + N., S. + V. . . . | 因為……;隨著…… |

For example:

1. With the departure of Eric Woods, Edith Ruby will be filling in for the time being.
2. With the next product release, we'll change our advertising strategy.

2. How to Conclude Announcements

+ In order to . . . , we are counting on the cooperation of all.
+ To facilitate . . ., the company expects everyone to get involved.
+ We are certain that with your . . ., the project will be a success.

下列是 Maples 台灣分公司的辦公室守則。

Employee Handbook

Our **philosophy**[1] is QUALITY. Above all else, we expect our employees to **uphold**[2] this principle. Speak, think, and write clearly and correctly. Do not forget to pay attention to details.

Please refer to the following company policies for **guidance**.[3] Speak to your supervisor if you are unclear on any of the points.

1. Working Hours

   1.1 Working hours are 8:30 a.m. to 5:30 p.m., Monday to Friday; lunchtime is 12:00 to 1:00 p.m.

   1.2 Break times will be scheduled **at the discretion of**[4] each employee's supervisor.

   1.3 Our office is open between 8:30 and 12:00 every Saturday morning. We post a schedule that gives everyone the opportunity to share the work, while still arranging whole weekends off on a **rotating**[5] basis. Please be advised that an important job may occasionally require one or more people to work all day.

2. Leaves of Absence

   2.1 Requests for time off should be made three days in advance to your supervisor (or the general manager if your supervisor is not around).

   2.2 Please note that if you take more than one day's sick leave in a row, a doctor's diagnosis is required.

   2.3 Other conditions are also set **in accordance with**[6] the Labor Standards Law.

3. Insurance

   Health, dental, and labor insurance all follow the Labor Standards Law. The company will also cover additional medical expenses that **result from**[7] injuries occurring on the job. Medical **reimbursement**[8] receipts are to be approved for payment by the general manager.

4. Salary

   4.1 Payday is on the last working day of each month.

   4.2 Your salary will be paid into your bank account.

   4.3 Accounting does not **withhold**[9] salary for tax purposes.

   4.4 Overtime will only be paid once your supervisor has signed for it.

5. Paid Holidays

   Paid holidays conform to the Labor Standards Law as follows:

   5.1 Seven days of paid vacation after one year of employment.

   5.2 10 days of paid vacation after three years of employment.

   5.3 14 days of paid vacation after five years of employment.

   5.4 The company will be offering 12 additional paid holidays this coming year. The holiday schedule has already been posted on the office bulletin board and is attached below as well.

HOLIDAY SCHEDULE FOR THE COMING YEAR	Lunar New Year — January 28 to February 2
	Peace Memorial Day — Tuesday, February 28
	Tomb Sweeping Day — Wednesday, April 5
	Labor Day — Monday, May 1
	Dragon Boat Festival — Wednesday, May 31
	Mid-Autumn Festival — Friday, October 6
	Double Tenth Day — Tuesday, October 10

◆ The information you **have access to**[10] is **confidential**.[11] **Be mindful of**[12] not repeating things you hear in the office.

中譯

員工手冊

我們的宗旨是「品質」。最重要的是我們期望員工們都能擁護此原則。清楚且正確地陳述、思考和書寫。切勿疏忽細節。

請參照下列公司政策以作為指導原則。如果對任何一點有不清楚的地方,可向主管提出。

1. 工作時間

1.1. 上班時間為星期一至星期五,早上八點半到下午五點半;午餐時間為十二點至下午一點。

1.2. 休息時間由主管自行安排。

1.3. 每週六早上八點半至十二點要上班。我們會公布輪值表,讓大家在分擔工作的同時,也能輪流安排週末假期。請注意有時會因有重要工作,而需一位或多位同仁上全天班。

2. 請假

2.1. 請假需於三天前向你的主管申請(如主管不在時,向總經理申請)。

2.2. 請注意連續請超過一天的病假,需檢附醫生診斷書。

2.3. 其餘情況皆遵照勞動基準法。

3. 保險

健康、牙醫和勞工保險皆遵循勞動基準法的規定。公司亦會負擔員工因公受傷的額外醫療費用。醫療賠償收據需經總經理核准使得支付。

4. 薪資

4.1. 發薪日為每月的最後一個上班日。

4.2. 薪資直接匯入你的銀行戶頭。

4.3. 會計部門不預扣稅款。

4.4. 需經主管簽核使得支領加班費。

5. 給薪假

給薪假規定遵照勞動基準法,如下:

5.1. 工作滿一年給予七天休假。

5.2. 工作滿三年給予十天休假。

5.3. 工作滿五年給予十四天休假。

5.4. 公司明年將提供另外十二天的給薪假。假期時間表已公布於辦公室的布告欄,並附於下方。

明年度假期時間表	農曆春節 — 一月二十八號至二月二號
	二二八和平紀念日 — 二月二十八號(二)
	掃墓節(清明節)— 四月五號(三)
	勞工節 — 五月一號(一)
	端午節 — 五月三十一號(三)
	中秋節 — 十月六號(五)
	雙十節 — 十月十號(二)

◆ 你所接觸到的資訊為機密。切勿將在辦公室聽到的消息散布出去。

Vocabulary and Phrases

1. **philosophy** [fə`lɑsəfɪ] *n.* 宗旨；看法；哲學

   I was asked about my philosophy towards work during the interview.
   我在面試時被問到我對工作的看法。

2. **uphold** [ʌp`hold] *v.* 擁護；支持

   The office does indeed uphold a dress code.
   辦公室的員工十分支持服裝規定。

3. **guidance** [`gaɪdn̩s] *n.* 指導；引導

   We always look to our more experienced coworkers for guidance.
   我們總是向經驗較豐富的同仁尋求指導。

4. **at the discretion of sb** 讓某人自行處理

   Yearly bonuses will be offered at the discretion of the office manager.
   年終獎金將由辦公室經理來決定是否提供。

5. **rotating** [`rotetɪŋ] *adj.* 輪流的

   Security guards in this building work the night shift on a rotating basis.
   這棟大樓的保全人員輪流值夜班。

6. **in accordance with** 依照

   In accordance with the contract, you must provide this product by next Tuesday.
   依照合約你必需於下週二前提供這項產品。

7. **result from** 起因於⋯⋯

   This eye problem resulted from sitting in front of a computer all day.
   這個眼睛的毛病起因於整天坐在電腦前。

8. **reimbursement** [.riɪm`bɜsmənt] *n.* 賠償；退款

   Will there be any reimbursement for travel expenses?
   差旅費可以報銷嗎？

9. **withhold** [wɪð`hold] *v.* 不給；保留；扣除

   We will withhold money from your paycheck and put it into the pension fund.
   我們將從你的薪資扣款，並存入退休基金。

10. **have access to** 接近；使用

    Nobody in our department will have access to your personal profile.
    我們部門沒有人能取得你的個人資料。

11. **confidential** [.kɑnfə`dɛnʃəl] *adj.* 機密的

    Please remember that anything we discuss during this meeting is confidential.
    請記得此次會議我們所討論的事項皆屬機密。

12. **be mindful of** 記住；小心

    Be mindful of how you talk to him. He has a bad temper.
    要注意和他説話的方式。他的脾氣不太好。

Language Corner

1. Following Rules

+ follow / obey
+ conform to
+ comply with
+ abide by
+ stick to
　遵從；信守

For example:

1. Failure to obey the regulations can result in a punishment.
2. All the new products need to conform to international safety standards.
3. The company strictly complies with the law.

2. How to Warn Someone

+ Be mindful of . . .
+ Try not to forget . . .
+ Please try to heed . . .
+ Remember that . . . is essential.
+ Please keep in mind that . . .

For example:

1. Remember that being punctual is essential.
2. Please keep in mind that we start work at nine o'clock sharp.

Unit Three
Liaison / Communication
聯絡

MAT 為 Maples 美國總公司在台灣的聯絡辦事處（liaison office），是總公司與客戶和製造工廠之間溝通的橋樑。

✔ Learning Goals

1. 計劃休假時，要指派職務代理人，並告知和自己業務有關的人

2. 無法代表公司和客戶討論技術問題時，應請求技術或品管人員的協助

3. 當業務往來的單位刻意要改變既有的運作模式時，需去函表達關切

Before We Start 實戰商務寫作句型

Topic 1: While You're Away . . . 不在時……

◆ I will be on [annual / maternity / funeral / sick . . .] leave from (date) to / until (date).
我將從（日期）休 [年 / 產 / 喪 / 病……] 假，休到（日期）。

◆ I will return to the office / get back to work / be at it again / be back on duty / be back on my post on (date).
我將在（日期）回來上班。

◆ Please direct any [questions / requests / concerns . . .] to my fill-in / replacement / substitute, while I'm away / on vacation / on holiday.
我不在 / 休假時，如果有任何 [問題 / 要求 / 疑慮……]，請找我的職務代理人。

◆ If [absolutely necessary / something urgent comes up / you're in a bind . . .], you can reach me on my cell phone / via e-mail.
如果 [真的有必要 / 有急事 / 你遇到困難……]，你可以打我手機 / 用電子郵件和我聯絡。

◆ I will be in touch with the office [periodically / on a regular basis / from time to time / now and then . . .].
我會 [定期 / 不時……] 與辦公室保持聯繫。

Topic 2: Showing Regret 表示遺憾

◆ I am / feel [sorry / regretful / disappointed / upset . . .] that . . .
我對……感到 [遺憾 / 失望 / 難過……]。

◆ It is unfortunate / regrettable / too bad that . . .
……令人感到遺憾。

◆ [Unfortunately / Regrettably / Sadly . . .], things have turned out . . .
[遺憾地 / 可惜地……]，結果是……

◆ There's nothing I can say about . . . except that . . .
除了……，我對……無話可說。

A. Advance Notice Before Taking Leave 休假通知

Maples 公司的營運總監 Mr. Woodruff 將要休年假，因此發信通知 OEM 工廠的業務經理。

To: David Yang
From: Earnest Woodruff
Date: March 13
Subject: Holiday Notice

Dear David,

 I will be on leave from Match 16th until March 24th, and will return to the office on March 25th. I will do my best to **clean up**[1] as many **pending**[2] issues with you as I can by the end of this week. While I am away, please **direct**[3] any questions or messages to my **fill-in**,[4] Ted Parker. Ted's extension number is 313. If absolutely necessary, he knows how to reach me. I will also be in touch with the office **periodically**.[5]

Regards,
Earnest

中譯

收件人：楊大衛
寄件人：恩尼斯特・伍德盧夫
日期：三月十三號
主旨：假期通知

親愛的大衛：

 我將從三月十六號休假到三月二十四號，並於三月二十五號回來上班。這個禮拜結束以前，我會盡量跟你處理完未解決的事項。我不在時你如果有任何問題或信息，請找我的職務代理人泰德・帕克，泰德的分機號碼是 313。如果真的有需要，他知道該如何聯絡我，我也會定期與辦公室聯繫。

祝 商安
恩尼斯特

Vocabulary and Phrases

1. **clean up** 處理；解決

Except for a few minor issues, I've basically cleaned this case up.
除了一些小問題，基本上我已經解決這個案子了。

2. **pending** [ˈpɛndɪŋ] *adj.* 懸而未決的

Many contract disputes with the supplier are still pending.
許多與供應商之間有關合約的爭議仍然沒有解決。

3. **direct** [dəˈrɛkt] *v.* 向……提出；將……指向

Please direct any questions you might have to Mr. Shannon.
如果你有任何問題，請找夏儂先生。

4. **fill-in** [ˈfɪl͵ɪn] *n.* 職務代理人；代替者

If you want to take a day off, you'll have to find someone to be your fill-in.
如果你要請一天假，你將必需找一個職務代理人。

5. **periodically** [͵pɪrɪˈɑdɪklɪ] *adv.* 定期地

I get in touch with our overseas clients periodically.
我定期與我們國外的客戶聯繫。

6. **in charge of** 負責；管理

The person in charge of this department is Miss Myers.
邁爾斯小姐負責這個部門。

B. Our Go-To Guy 聯繫窗口

有日本客戶問到有關技術方面的問題，行銷經理 **Mr. Strong** 請工程部代為回覆。

To: Mr. Asahina
From: Scott Simmons
Date: March 2
Subject: PC-1 Technical and QC Matters

Dear Mr. Asahina,

Mr. Strong has forwarded to me your recent fax regarding the technical / QC matters you wish to discuss. Mr. Strong is **in charge of**[6] our marketing department, and he does not normally represent Maples in technical meetings. If it's technical **expertise**[7] that you need, I'll be more than willing to answer any questions. I will be in Tokyo on business for about one week. If you like, we could set up a meeting to **run over**[8] any further details.

Please let me know what you would like to do.

Yours truly,
Scott Simmons
Vice President of Engineering
Maples Taiwan

中譯

收件人：朝比奈先生
寄件人：史考特‧希蒙斯
日期：三月二號
主旨：PC-1 技術與品管問題

親愛的朝比奈先生：

斯壯先生已將您最近傳的傳真轉寄給我，上頭有您欲討論的技術與品管事宜。斯壯先生主管我們的行銷部，他通常不會代表 Maples 公司出席有關技術議題的會議。如果您需要技術方面的專業知識，我很樂意回答任何問題。我將去東京出差約一個星期，如果您願意的話，我們可以安排會面，討論進一步的細節。

請讓我知道您的決定。

謹啟
史考特‧希蒙斯
工程部副總裁
Maples 台灣分公司

---

7. expertise [ˌɛkspɚˈtiz] *n.*
專業知識（技術）

His area of expertise is computers.
他的專業領域是電腦。

8. run over 討論；解釋

Could you run over the incentive package one more time, please?
可以請你再解釋一次那個獎勵方案嗎？

Language Corner

1. Finishing Up

I will do my best to clean up as many pending issues as I can by the end of this week.

You Might also Say . . .

- I will try my hardest to finish up all the pending work in the next few days.

- I will be sure to get this all sorted out before week's end.

- I promise I'll have this case wrapped up before I go on vacation.

2. Adding Stress （強調句型）

If it is . . . that S. + V., S. + V.
如果是……，那……

For example:

1. If it is the shipping date that you're not happy about, we can try to do something about it.

2. If it was Mr. Douglas who edited this report, then I need to talk to him.

註：如果強調的部分為「人」，亦可用 who 代替 that。

C. Operating Procedures 運作模式

Maples Taiwan 去函給 OEM 工廠，希望對方不要改變現有的運作模式。

To: David Yang
From: Richard Barlow
Date: September 27
Subject: Your Fax #S-723

Dear Mr. Yang,

I'm confused and disappointed by the **tone**[1] of your fax. To tell you the truth, this really **caught me off guard**,[2] but I'll try to answer as best I can.

I was informed that you have stopped giving us copies of your QC inspection reports. This is **unfortunate**[3] because I believe we can best control the quality of GasPack® products by sharing information. There's nothing I can say about your decision, Mr. Yang, except that I hope you'll **reconsider**.[4] I'm really sorry that things have turned out this way.

Plus, I don't seem to understand the second item in your message. It sounds as if you'd like to change some operating procedures that are currently **in place**.[5] Maybe it would be a better idea if we met in person **as opposed to**[6] carrying this issue out **via**[7] e-mail.

I hope you can get back to me as soon as possible.

Regards,
Richard

中譯

收件人：楊大衛
寄件人：理查‧巴洛
日期：九月二十七號
主旨：你編號 S-723 的傳真

親愛的楊先生：

你傳真的語氣讓我感到困惑與失望。老實說，我真的很驚訝，但我還是會儘量回答你的問題。

我被告知你已經停止提供我們你們品管檢驗報告的副本，這點令人感到遺憾，因為我相信藉由資訊共享，才最能控管 GasPack® 產品的品質。楊先生，除了希望你能重新考慮外，對於你的決定我無話可說。我非常遺憾事情演變成這樣。

另外，你傳真中提到的第二點我也不太能理解。聽起來你好像想改變一些現行的運作模式。相對於透過電子郵件來解決這個問題，也許我們見面談會比較好。

我希望你能儘快回覆我。

祝 商安
理查

36

Vocabulary and Phrases

1. **tone** [ton] *n.* 語氣；腔調

   The tone of her e-mail was quite friendly.
   她那封電子郵件的語氣相當和善。

2. **catch sb off guard** 乘某人不備；使某人措手不及

   News of the merger caught us off guard.
   公司合併的消息讓我們感到措手不及。

3. **unfortunate** [ʌnˋfɔrtʃənɪt] *adj.*

   令人遺憾的；不幸的

   It's unfortunate that Tina's going to be
   transferred. We'll really miss her.
   令人遺憾的，蒂娜即將要被調職了。我們會很想念她的。

4. **reconsider** [ˌrikənˋsɪdɚ] *v.* 重新考慮

   I hope you'll reconsider our offer.
   我希望您能重新考慮我們的報價。

5. **in place** 現行的；已制定的

   Please note that the company has several
   new regulations in place.
   請注意公司已制定數項新規定。

6. **as opposed to** 相對於……；而非……

   Mr. Rota wants to hold the meeting this
   weekend as opposed to next Monday.
   羅塔先生想要在本週末舉行會議而非下週一。

7. **via** [ˋvaɪə] *prep.* 透過；經由

   The entire conversation was held via
   conference call.
   整個談話的進行都是透過電話會議。

🌐：® 和 ™ 是常見的商標符號，前者表示「已註冊過的
商標（registered trademark）」，後者則指「一
般的商標（trademark），不一定有註冊過」。

Language Corner

1. Expressing Surprise

This really caught me off guard.

You Might also Say . . .

- This came right out of the blue.
- I can't believe this. I never even saw it coming.

2. Asking Someone to Reconsider

There's nothing I can say about your decision
except that I hope you'll reconsider.

You Might also Say . . .

- Is there any way I can change your mind?
- We hope that's not your final decision.
- How can we persuade you to reconsider?

Maples Taiwan 的負責人 Richard 向美國總公司的總裁 Tony 報告一些雜事。

To: Tony Hughes
From: Richard Barlow
Date: March 25
Subject: Office Matters

1. New **Letterhead**[1]

We're expecting samples of the new letterhead this Friday. If we approve it, then I'll type out the **profile**[2] on that new letterhead. When we get something we think is ready for the printer, I'll make sure you see it **beforehand**.[3]

2. Schedule

We're looking forward to your visit. Maybe you can meet up with Bob Sanders, who is also planning a trip around that time. Before you **finalize**[4] your **itinerary**,[5] please be advised that Mica will be closed on the following dates for various reasons:

April 3, 4, 5, 9, 10, 11, 12

The 4th and 5th are national holidays; the rest of the days are Mica holidays only.

3. Jack and I will be out of the office tomorrow, so you won't get anything **in-depth**[6] through e-mail or fax. We will, however, be **keeping up with**[7] things at the office.

Take it easy,
Richard

中譯

收件人：東尼・休斯
寄件人：理查・巴洛
日期：三月二十五號
主旨：辦公室事務

1. 新的公司信紙

我們預計本週五可以看到新的公司信紙樣本，如果我們核准新的信紙，我會把簡介打在上面。當我們準備好我們認為可以送印的樣本時，我會確保先讓您看過。

2. 時間表

我們很期待您的來訪，或許您可以和鮑伯・桑德斯碰面，他也計畫在那個時間過來。在您最後確定您的行程前，請注意由於種種因素 Mica 工廠將在下列日期關閉：

四月三號、四號、五號、九號、十號、十一號、十二號

四號和五號是國定假日，其餘則是 Mica 工廠內部的休假日。

3. 我和傑克明天不在辦公室，所以您無法透過電子郵件或傳真得到詳細的回覆，不過我們還是會隨時掌握辦公室的最新狀況。

放輕鬆
理查

Vocabulary and Phrases

1. **letterhead** [`lɛtɚ͵hɛd] *n.* 印在信紙的信頭；印有信頭的信紙

   You'll see the address on the letterhead.
   你可以在信頭上看到地址。

2. **profile** [`pro͵faɪl] *n.* 簡介；概況

   Could you please provide a short profile introducing your company?
   可以請你提供你們公司的簡介嗎？

3. **beforehand** [bɪ`for͵hænd] *adv.* 事先；提前地

   We promise you can read the press release beforehand.
   我們保證會事先讓您看新聞稿。

4. **finalize** [`faɪn͵laɪz] *v.* 敲定；完成

   Ms. Wang is flying in tomorrow to finalize the terms of the contract.
   王女士明天將飛過來敲定契約的條件。

5. **itinerary** [aɪ`tɪnə͵rɛrɪ] *n.* 旅程；旅行計畫

   Here's the itinerary for your trip to Cambodia.
   這是你去柬埔寨的行程。

6. **in-depth** [`ɪn`dɛpθ] *adj.* 深入的；徹底的

   The magazine wants to conduct an in-depth interview with our CEO.
   那本雜誌想和我們的執行長做一個深入的訪談。

7. **keep up with** 了解……的最新狀況

   Are you keeping up with the latest on Wall Street?
   你了解華爾街的最新狀況嗎？

Language Corner

Staying Informed

We will be keeping up with things at the office.

You Might also Say . . .

- We will be on top of what's going on at the office.
- We will keep track of what's happening in the office.
- We will keep / stay abreast of the latest developments at the office.

Insight

meet / meet (up) with 用法大不同

meet 可指「認識；初次見面」或「與（熟人）碰面」，而 **meet (up) with** 通常指「和（某人）約好見面；和（某人）開會」。

For example:

1. I still haven't met Mr. Johnson's new assistant.
2. He's meeting his wife for lunch.
3. If you have time, let's meet with Rob Darcy and work more on the proposal.
4. They will be meeting up with some clients Friday afternoon.

Unit Four
Financial Matters
財務

MAT 的營運開銷完全仰賴美國總公司的匯款，因此匯款時機及金額多寡是雙方聯繫的重點，如因溝通不良而導致資金無法及時匯入，會影響到辦事處的運作。

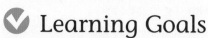 Learning Goals

1. 如何減少辦公室的開銷，以免超出預算

2. 員工提案獎勵制度的實施可有效幫公司節流

3. 針對新的會計年度，訂出更嚴謹的會計作帳系統

Before We Start 實戰商務寫作句型

Topic 1: Talking about Expenses 談論開銷

* Most [categories / estimates / figures . . .] are within our budget goals / within (on) budget / on target.
  大部分的 [類目 / 預估 / 數字……] 都在我們的預算目標之內。

* One area / place / section in which we overspent was [marketing / advertising . . .].
  我們超支的部分是 [行銷 / 廣告……]。

* I'd like to ask (for) your help in [reducing expenses / cutting costs / promoting efficiency / getting this message across . . .] even more.
  我要請你幫忙再更進一步（再多加）[減少開支 / 降低成本 / 提升效率 / 散布這則訊息……]。

Topic 2: Talking about Problems 談論問題

* There is a problem / an issue with . . .
  ……有問題。

* I'm having a hard time / trouble / difficulty + V-ing . . .
  我在……方面有困難。

* [Missing a payroll / Delaying the money transfer . . .] was a terrible / serious / huge / fatal mistake / error.
  [未支付薪資 / 延遲匯款……] 是很嚴重 / 無可挽回的錯誤。

* I don't think that . . . is a viable option / alternative / proposition.
  我不認為……是可行的選擇 / 提議。

* I'm sorry to lay this on you / inconvenience you / cause you trouble . . .
  不好意思麻煩你……

* . . . has once again hit (run into) a snag / hit a bad patch / encountered problems.
  ……又有問題了。

A. Budgetary Control 預算控制

Maples 台灣分公司的負責人 Richard 寫信給員工，說明預算執行的情形，並要求減少雜項支出。

January 17
To: ALL MAT EMPLOYEES
From: Richard Barlow

I've just reviewed our **expenditures**[1] for the first quarter of the fiscal year.* I'm happy to report that most **categories**[2] were within our budget goals.

One area in which we **overspent**,[3] however, was Miscellaneous Office Expenses. This category includes phones, faxes, stationery, shipping, postage, and so on. I'd like to ask your help in reducing this area of expenses even more.

Here are a few suggestions:

1. Communicate by e-mail as much as possible.
2. Make international calls after 4:00 p.m. Remember that the most expensive time to make international calls is between 8:00 a.m. and 4:00 p.m.
3. We are **charged**[4] for local calls once a monthly call **allowance**[5] is **exceeded**[6] on any of the four phone lines. So, for local calls, use these phone lines on an equal basis. This is important because we are billed separately for each line.
4. If you can think of ways to save money in MAT operations, write them down along with calculations indicating how much money they will save. If your calculations are confirmed and your suggestion **implemented**,[7] you will receive a cash bonus of 50 percent of the **anticipated**[8] first-year savings as part of our employee suggestion bonus program.

Thanks for your cooperation.

Richard

中譯 一月十七號
收件人：MAT 全體員工
寄件人：理查 · 巴洛

我剛審查完我們會計年度第一季的支出。我很高興向各位報告大部分的類目都在我們的預算目標之內。

不過我們在「辦公室雜項支出」的部分超支了。這個類目包含電話、傳真、文具、運費、郵資等。我要請你們幫忙再多減少一些這方面的開支。

這裡有一些建議：

1. 儘可能使用電子郵件來聯繫。

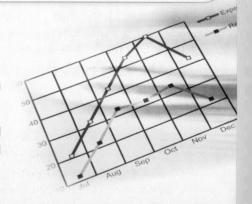

2. 下午四點之後再打國際電話。記住撥打國際電話最貴的時段為早上八點至下午四點。

3. 四支電話線其中一支超過每月基本通話時數,我們就要支付本地通話費。所以請平均使用這四線。這很重要,因為這四支電話線我們是分別被計費的。

4. 如果你能想到節省 MAT 營運開銷的方法,請將辦法寫下來並估算可以節省多少費用。依照員工提案獎勵制度,如果你的估算被證實無誤且建議被採用,你將可以獲得預期第一年所省下經費百分之五十的現金獎金。

感謝你們的合作。

理查

Vocabulary and Phrases

1. **expenditure** [ɪk`spɛndɪtʃə] *n.* 支出;經費

   Please submit any travel-related expenditure within two weeks of return from travel.

   請於出差回來的兩週內提交相關旅費申請。

2. **category** [`kætə‚gɔrɪ] *n.* 類別;種類

   Please take a look at the third category from the bottom.

   請看一下倒數第三個類別。

3. **overspend** [‚ovə`spɛnd] *v.* 超支

   The department has overspent its budget by NT$100,000.

   該部門的預算已經超支新台幣十萬元了。

4. **charge** [tʃɑrdʒ] *v.* 索價;對……收費

   If you lose your employee swipe card, you'll be charged NT$100 for a new one.

   如果你的員工門禁卡遺失了,你要付新台幣一百元以取得新卡。

5. **allowance** [ə`lauəns] *n.* 分配額;限額

   If your luggage surpasses the weight allowance, you'll be charged extra.

   如果你的行李超過重量限制,你要多付費。

6. **exceed** [ɪk`sid] *v.* 超過(限度、範圍)

   Your department's performance has exceeded our wildest expectations.

   你們部門的表現遠遠超過我們的預期。

7. **implement** [`ɪmpləmənt] *v.* 實施;執行

   The changes we suggested still haven't been implemented.

   我們所建議的改變仍然未被實施。

8. **anticipated** [æn`tɪsə‚petɪd] *adj.* 預期的

   The boss wants to know the anticipated outcomes.

   老闆想要知道預期的結果。

Biz Focus

★ fiscal year　會計年度;財政年度

Language Corner

Trying Your Best

| as + adv. + as possible / one can | 儘量…… |

For example:

1. I need you to take care of this as soon as possible.
2. We need to reduce expenses as much as possible.
3. The client is waiting at the airport. I need you to get out there as fast as you can.
4. The company will implement these new hiring policies as soon as it can.

B. Request for a Fund Transfer 要求匯款

#1 Richard 要求美國總公司立即匯款，以支付台灣分公司員工的薪資。

To: Ted Parker
From: Richard Barlow
Date: March 31
Subject: Missing Payroll

Dear Ted,

Yesterday, you asked if there is a problem with this month's transfer. To be frank, there is. Because the transfer was both late and not **in full**,[2] I couldn't pay my people today. Monica also has to go down to the phone company and pay the phone bill **in person**[3] this afternoon because they are going to cut off our service.

As our employees are afraid they're not going to get paid at the end of the month, or that the company is going to **fold**[4] any day, I'm having a hard time keeping them **motivated**.[5] And if we aren't closing the doors, then missing the payroll was a terrible mistake. I'm sure you sent too little too late because you're having cash flow* problems, but I don't think that missing our payroll is a **viable**[6] option.

In the future, I'll be sure to check on the transfer status earlier in the month. And maybe next time there's a cash flow problem, MCUS could wait for their salary and MAT could get paid on time. I'm sorry if this sounds **harsh**.[7] I'm just trying to give you a **straightforward**[8] account of what's going on.

Sincerely,
Richard

中譯 #1

收件人：泰德・帕克
寄件人：理查・巴洛
日期：三月三十一號
主旨：未支付薪資

親愛的泰德：

你昨天問我這個月的匯款是否有問題。老實說，的確有問題。因為匯款延遲且只匯了一部分，導致我今天無法支付薪資給我的員工。今天下午莫妮卡還必需親自去電信公司繳電話費，因為他們將要停止提供我們服務。

由於員工擔心月底會領不到薪水，或是公司隨時會倒閉，我很難激勵他們。而如果我們沒有要結束營業，那麼未能支付薪資便是一個很嚴重的錯誤。我確信你會延遲匯款且匯款的金額不足，是因為你的現金流轉出了問題，但我認為選擇不支付我們的薪資是不可行的。

以後每個月我一定會提前確認匯款的狀況。或許下次遇到現金流轉的問題時，Maples 美國總公司的薪資可以延後發放，而讓台灣分公司的員工可以準時領薪水。如果這些話聽起來很刺耳的話，我感到很抱歉。我只是試著坦率地向你說明實際的狀況。

祝 商安

理查

Vocabulary and Phrases

1. **transfer** [`trænsˏfɚ] *n.* 匯款
   We usually make transfers by wire.
   我們通常使用電匯。

2. **in full** 全部；全數
   I want to be paid in full by the end of the month.
   我想要在月底前領足全部薪水。

3. **in person** 親自
   You can only apply for a VIP card in person.
   你只能親自申請貴賓卡。

4. **fold** [fold] *v.* （事業等）失敗；關閉
   More than 4,000 people lost their jobs when the company folded.
   當該公司倒閉時，超過四千人失業。

5. **motivated** [`motəˏvetɪd] *adj.*
   有積極性的；有動機的
   We pay our employees a good wage to keep them motivated.
   我們付給員工優渥的薪資以使他們保有積極性。

6. **viable** [`vaɪəbḷ] *adj.* （計畫等）可行的；可實施的
   An Asian style of management may not be viable in the West.
   亞洲的管理作風在西方不一定可行。

7. **harsh** [hɑrʃ] *adj.* 刺耳的；嚴厲的
   Maybe you were too harsh with Tina. She looks like she's going to cry.
   或許你對蒂娜太嚴厲了。她看起來快哭了。

8. **straightforward** [ˏstret`fɔrwɚd] *adj.*
   坦率的；直接的
   The reason I like Hank is that he is straightforward.
   我喜歡漢克是因為他很直率。

Biz Focus

★ cash flow （會計）現金流轉；現金流量

Language Corner

1. Making Your Point

To be frank, S. + V.
To tell (you) the truth, S. + V.
Frankly / Honestly speaking, S. + V.

老實說……

For example:

1. To be frank, that's an ugly tie.

2. To tell you the truth, I'm not very happy with his performance.

2. Don't Take It Personally

I'm sorry if this sounds harsh.

You Might also Say . . .

- I hope you don't take this badly / feel offended.

- This might be a little hard to swallow.

#2 Richard 遲遲沒收到美國總公司的匯款，於是傳真給在日本出差的董事長，希望他介入協調。

Please pass this message to Mr. Tony Hughes of Maples Corp., who will be checking into your hotel today. Thank you.

From: Richard Barlow
Fax Number: 886-2-2788-5113
Oct. 22   06:10 PM   PAGE 1 OF 1

Dear Tony,

I hope you're having a good trip. I'm sorry to lay this on you, but it can't wait.

The money transfer has once again **hit a snag**.[1] This is the second **consecutive**[2] month that I will be forced to either make payroll out of my own pocket or tell my folks there's no money to pay them. In spite of the letters I sent Ted and Shelly **reminding** them **of**[3] the transfer, nothing was done. Even worse, I received a request today, after the money should have already been sent, for a **breakdown**[4] of the funds required. The transfer schedule has been basically unchanged for the past three years. To tell you the truth, I'm a bit **peeved**.[5] I think we need to implement a system to transfer funds **reliably**[6] so that I don't have to chase Ted and Shelly around. I would really appreciate your **intervention**[7] in this situation.

By the way, good luck with the Japan **talks**.[8]

Best regards,
Richard

中譯 #2

請將這封訊息交給 Maples 公司的東尼・休斯先生，他今天將會入住貴飯店。謝謝你。

寄件人：理查・巴洛
傳真號碼：886-2-2788-5113
十月二十二號　　　下午六點十分　　　第一頁，共一頁

親愛的東尼：

我希望您這趟旅程愉快平安。很抱歉麻煩您，但此事刻不容緩。

匯款又出問題。這已經是連續第二個月我要不是被迫拿自己的錢來付薪水，就是要告知我的員工沒有錢去支付他們。儘管我已寄信提醒泰德和雪莉此筆匯款，他們仍未有任何作為。更糟的是，在過了早該匯款的時間點之後，我還被要求提供所需資金的明細。基本上過去三年匯款時間表都沒有改變。老實説，我有點生氣。我想我們需要實行一個能確實匯款的制度，如此一來我就不用追著泰德和雪莉跑。如果您能介入協調此事，我將非常感激。

順帶一提，祝您在日本的會談一切順利。

謹上
理查

Vocabulary and Phrases

1. hit a snag 遇到突如其來的困難

They hit a snag while negotiating the contract.
他們在協商合約時遇到突如其來的困難。

2. consecutive [kən'sɛkjətɪv] *adj.*
連續的；接連的

She has worked on this project for three consecutive days.
她已經連續處理這個案子三天了。

3. remind sb of sth 提醒某人某事

Remind me of this issue first thing tomorrow.
明天一早就提醒我這個問題。

4. breakdown ['brek͵daun] *n.* 明細；分類

During the presentation, Art gave a breakdown of the numbers.
簡報時亞特提供了那些數據的明細。

5. peeved [pivd] *adj.* 惱怒的

Whenever the manager sees us fooling around in the office, she gets peeved.
每當經理看到我們在辦公室內鬧晃時，她就會生氣。

6. reliably [rɪ'laɪəblɪ] *adv.* 確實地；可靠地

Don't worry. Debbie always behaves reliably.
別擔心。黛比做事向來很可靠。

7. intervention [͵ɪntɚ'vɛnʃən] *n.* 介入；調停

Without the government's intervention, the bank could have gone out of business.
如果沒有政府的介入，那家銀行可能早就關門了。

8. talks [tɔks] *n.* 正式會談；談判
（作此義解時，常用複數）

Talks are under way to avert a hostile takeover.
談判正在進行以防止惡性收購。

Language Corner

1. One After Another

This is the second consecutive month . . .

You Might also Say . . .

- This is the second month in a row . . .
- This is the second month in succession . . .
- This is the second straight month . . .

2. It's Not What We Thought

In spite of / Despite + N. / V-ing, S. + V. . . .
In spite of / Despite the fact (that) S. + V. . . ., S. + V. . . .

儘管……

For example:

1. In spite of the weather, the plane took off.
2. In spite of having a GPS navigation device in his cab, the driver still couldn't find the office.
3. Despite the fact that Heath talked on the phone all afternoon, he was still able to finish the project.

C. Expenditure Breakdown 支出細目

Richard 列出明年度的主要支出，並與財務部討論匯款流程。

To: Ted Parker
From: Richard Barlow
Date: 8/2
Subject: Money Matters

1. I don't know if Tony discussed this with you, but he and I have agreed on new housing terms. The house **rental**[1] is US$900 per month, **payable**[2] in one **lump sum**[3] before the year commences. I have already prepaid the required two months' **deposit**[4] of US$1,800.

2. I'm sure you can appreciate that we must pay the rent on the agreed-upon date. Therefore, I hope you will be able to send these funds early. Please remember that we can only **withdraw**[5] US$5,000 per day from our account. Often we need several days to withdraw the necessary amount to make payroll and pay the **assorted**[6] bills. The money for the deposit and air conditioners isn't as **time sensitive**,[7] so it can be delayed.

3. We need to figure out how we're going to handle the coming fiscal year. We might want a slightly different transfer and / or accounting system. We also have to tighten up the current system. For example, I'd like to get copies of my monthly company credit card **statements**[8] so I can track my budget more closely. It's been **hit-or-miss**[9] until now. Each month, I will submit a request for the amount of cash I need transferred for the following month so we are not running short some months and sitting fat other months anymore.

Thanks and best regards,
Richard

中譯 收件人：泰德‧帕克

寄件人：理查‧巴洛

日期：八月二號

主旨：財務事項

1. 我不知道東尼是否和你討論過，但他和我都同意新的租屋條款。辦公室的租金是每個月九百元美金，在年度開始之前必需一次付清。我已經依要求預付了兩個月的保證金一千八百元美金。

2. 我相信你能諒解我們必需在議定的日期支付租金。因此我希望你可以提前匯出這些款項。請記住我們每日只能從我們的帳戶中提領五千元美金。通常我們需要好幾天來提領需要的金額以支付薪資和各種帳單。而保證金和空調的費用並沒有時間性，所以可以延遲。

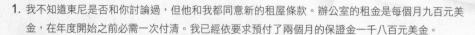

3. 我們需想想如何處理即將來臨的會計年度。我們可能會需要稍微不同的匯款方式和（或）會計制度。我們也要更加嚴格控管現行的制度。舉例來說，我想要取得我每個月公司信用卡的帳單，好讓我可以更嚴密地追蹤我的預算，這直到現在都是漫無計畫的。以後每個月我將會提交一份月所需匯款的資金總額需求，這樣我們的資金就不會再有幾個月短缺、幾個月又過剩的情形。

謝謝 謹上

理查

Vocabulary and Phrases

1. **rental** [ˈrɛntl̩] *n.* 租金
   The office rental is going up again.
   辦公室的租金又漲價了。

2. **payable** [ˈpeəbl̩] *adj.* 應支付的；到期的
   The interest is payable twice a year.
   利息需一年支付兩次。

3. **lump sum** 一次付款額；總金額
   Mike wants to pay off his mortgage in one lump sum.
   麥克想要一次付清他的房貸。

4. **deposit** [dɪˈpɑzɪt] *n.* 訂金；存款
   We have already put a big deposit down on that land.
   那塊土地我們已經支付了一大筆保證金。

5. **withdraw** [wɪðˈdrɔ] *v.* 提領；取消
   The bank teller told me that I wouldn't be able to withdraw the funds until after the weekend.
   銀行出納員告訴我過了週末我才能提款。

6. **assorted** [əˈsɔrtɪd] *adj.* 各種的
   We'd like to order 50 boxes of assorted Belgian chocolates.
   我們想要訂購五十盒綜合的比利時巧克力。

7. **time sensitive** 有時間性的
   Since it's time sensitive, send it by express mail.
   因為這是有時效性的，所以要寄快遞。

8. **statement** [ˈstetmənt] *n.* 結算單；報表
   The shareholders will pay close attention to our annual financial statements.
   股東們將會密切關注我們的年度財務報表。

9. **hit-or-miss** [ˌhɪtɚˈmɪs] *adj.* 無計劃的；隨意的
   With Andrew on vacation, things have been hit-or-miss around the office.
   因為安德魯休假，整個辦公室也跟著失了序。

Language Corner

Making Adjustments

We have to tighten up the current system.

You Might also Say . . .

- We have to make the current system more efficient.
- We should revamp the current system.
- The current system needs to be sorted out.

Chapter Two

Agents / Agencies

Unit One
Business Plan 營運計畫書

Richard Barlow 來台工作幾年之後，除了原先負責的 Maples Taiwan，又打算成立 Far East Technical Services 這家公司，替一些外國公司做在台採購及品管檢驗的工作。這篇文章就是 Richard 為了成立新公司，而寫給美國總公司的對內創業計畫書。

✔ Learning Goals

1. 說明創業目的及未來營運目標

2. 評估是否具有競爭優勢

3. 說明市場現況、商機及未來發展

4. 明確指出利基所在

5. 計畫書力求簡單明瞭，切勿過於冗長或使用太多專業術語

Before We Start 實戰商務寫作句型

Topic 1: Including Something Extra 附上資料

* For your review / consideration, enclosed / attached is . . .
  附件為……供您審閱 / 參考。

* I've included / attached / added . . .
  我附上……

* Please find enclosed / affixed . . .
  請參考附加的……

Topic 2: How Things Are Coming Along 說明事情進展

* I am [planning / attempting / preparing / making arrangements . . .] to V. . . .
  我 [打算 / 試圖 / 準備……] ……

* We are in the process / middle / course of + V-ing . . .
  我們正在……

* Meanwhile / At the same time / Concurrently, S. + V. . . .
  同時……

* At this point / At the present moment / (Up) until now, S. + V. . . .
  （直到）此刻……

Topic 3: Forecasting 預測

* . . . could generate [commissions / profits . . .] of (amount of money / percentage) if it (they) materialize(s) / come(s) into effect / come(s) into play.
  如果……實現的話，可帶來（金額 / 百分之幾）的 [佣金 / 利潤……]。

* If we were to extrapolate / project / expand this over the next few years, it would [paint a very rosy picture / show some good results . . .].
  如果我們以此來推斷往後幾年的情形，[前景十分看好 / 會有一些不錯的成果……]。

* I cannot offer [a valid dollar assessment / an accurate estimate / a meaningful projection / conclusive figures . . .] right now.
  我目前無法提供 [有根據的金額評估 / 正確的估計 / 有意義的預測 / 確實的數據……]。

TO: Tony Hughes
FROM: Richard Barlow
DATE: November 21
SUBJECT: FETS Business Plan
ATTACHMENT: FETS Briefing

Dear Tony,

For your review, enclosed is my business plan. Please let me know if there's any other information you would like.

Sincerely,
Richard

FAR EAST TECHNICAL SERVICES BUSINESS PLAN
BY RICHARD BARLOW

EXECUTIVE SUMMARY:

I am planning to establish a technical service business in Taiwan, one that offers the staff of Far East Technical Services as experts in the areas of manufacturing, sourcing,* and quality control. Currently, the **strengths**[1] of the FETS staff are in the engineering and quality-control areas, but the **bulk**[2] of the work that's come through our door so far indicates a stronger market for sourcing than manufacturing and QC. We are in the process of evaluating our strengths and weaknesses, and adjusting the organization accordingly. Meanwhile, we plan to **court**[3] more opportunities and remain open to different ideas. At this point, we do not plan to **curb**[4] the **scope**[5] of our activities in any way.

MARKET ANALYSIS:

Our target customers are small- to medium-sized companies, usually located in but not limited to the Western Hemisphere. We have advertised on a smaller **scale**[6] with satisfactory results, and plan to expand on this. We've received numerous **inquiries**[7] from personal contacts, and expect the trend will continue.

中譯　收件人：東尼・休斯
寄件人：理查・巴洛
日期：十一月二十一號
主旨：FETS 營運計畫書
附件：FETS 簡報

親愛的東尼：

附件為我的營運計畫書，供您審閱。假如您還需要其他的資訊，請讓我知道。

祝 商安
理查

FAR EAST TECHNICAL SERVICES 營運計畫書
理查・巴洛 撰

執行摘要：

　　我正打算在台灣成立一家技術服務公司，讓 Far East Technical Services 的員工變成製造、採購和品管等領域的專業人員。現在 FETS 員工的優勢是在技術和品管領域，但是到目前為止大部分上門的生意顯示在採購方面的市場需求大於對製造和品管方面的需求。我們正在評估自己的優劣勢，並依此調整公司的編制。同時，我們計劃爭取更多機會並廣納不同的意見。此刻我們不打算以任何方式來限制我們的業務範疇。

市場分析：

　　我們的目標顧客群是中小型企業，通常是但並不侷限於西半球國家。我們打過小規模的廣告且得到令人滿意的結果，我們也計劃再擴大廣告規模。透過私人的關係，我們已經接到許多客戶的詢問，並預期這種情形會持續下去。

▶ Vocabulary and Phrases

1. strength [strɛŋθ] *n.* 優勢；強項（weakness *ant.*）

The interviewee was asked to list her strengths.
那位面試者被要求列出她的強項。

2. bulk [bʌlk] *n.* 大部分；大多數

The bulk of our orders come from the United States.
我們大多數的訂單都來自美國。

3. court [kɔrt] *v.* 企圖獲得；追求

In order to expand, we started courting business in other countries.
為了擴展，我們開始到其他國家爭取生意。

4. curb [kɝb] *v.* 抑制；遏止

Factories will be required to curb their emissions.
工廠將被要求抑制其所排放的廢物。

5. scope [skop] *n.* 範圍；領域

The scope of his vision far exceeded anything our manager had ever seen.
他的視野遠超過我們經理。

6. scale [skel] *n.* 規模；大小

The company wants to take on projects of a larger scale in the future.
該公司未來想承接規模更大的案子。

7. inquiry [ɪnˋkwaɪrɪ] *n.* 詢問；打聽
（英式英文拼法為 enquiry）

Please direct any inquiries you might have to our customer service department.
如果您有任何問題，請找我們的客服部。

▶ Biz Focus

★ **sourcing** 尋找供應商

　　負責 sourcing 的人會先取得幾家可能供應商的報價，並會同品管、技術或業務等部門的代表一起訪查、評估，選定供應商後，再由採購人員執行後續採購的工作。

CURRENT OPERATIONS:

We have established a working relationship with AECO, and have even become their Taiwan office. We shipped our first commission order* last month in cooperation with them. In all, we are working on five or six AECO projects. Recently, an engineering service company in New Jersey called TEK Industries began giving us projects to **quote**.[8] We have a quote out to a Dutch company to build gearmotors* worth an initial order of US$100K. There will be **substantial**[9] follow-up orders as well. Finally, an American baby furniture company is considering a contract proposal worth US$60K per year in retainer fees.*

Since we began FETS, we've offered quotations to the following companies:

- Pacific Taximeter, New Hampshire
- MicroSensation Systems, Ohio
- Kay Industrial BV,* Netherlands
- Senture Products, Massachusetts
- AECO, New York (several branches)

We have also made personal contacts with the following companies. They have all shown an interest in pursuing a relationship with FETS in the future:

- AD Associates, New Hampshire
- CPS Enterprises, Canada
- LG Toys, California
- Roy Mack
- TEK Industries, New Jersey
- Walker Power Products, California

FORECAST:

At first glance, FETS looks to have a quick **ramp-up**[10] to healthy business targets. Right now, we have quotations out that could generate commissions of US$50K to 100K if they all **materialize**.[11] Our level of activity is increasing steadily, and if we were to **extrapolate**[12] the activity of the first two months over the next 10 months, it would paint a very **rosy**[13] picture. Start-ups can be difficult to assess; therefore, I cannot offer a **valid**[14] dollar assessment at the moment.

中譯

目前營運：

    我們已和 AECO 建立了合作關係，甚至成為他們在台灣的辦事處。上個月我們一起合作出了第一筆抽佣式的訂單。我們一共處理五、六個 AECO 的案子。最近一家位在紐澤西名為 TEK Industries 的技術服務公司，也開始將案子交給我們報價。我們已向荷蘭一家公司提供齒輪馬達的報價，第一批訂單價值為十萬元美

56

金，後續還會有更多的訂單。最後是一間美國嬰兒傢俱公司正在考慮我們的一份合約提案，每年會有六萬元美金的服務費。

從我們成立 FETS 以來，我們已為下列公司提供報價服務：

- Pacific Taximeter（新罕布夏州）
- MicroSensation Systems（俄亥俄州）
- Kay Industrial 私人有限公司（荷蘭）
- Senture Products（麻州）
- AECO（紐約）（數間分公司）

我們也和下列幾間公司接觸過。他們都有興趣在未來與 FETS 建立合作關係：

- AD Associates（新罕布夏州）
- CPS Enterprises（加拿大）
- LG Toys（加州）
- Roy Mack
- TEK Industries（紐澤西州）
- Walker Power Products（加州）

預測：

乍看之下，FETS 似乎快速成長並將達到極佳的業績目標。目前的報價都談成的話，可帶來五萬到十萬元美金的佣金。我們的業務量穩定成長，而如果我們以頭兩個月的業績來推斷接下來十個月的業務量，前景十分看好。但公司剛成立很難去做評估，所以我目前無法提供有根據的金額預估。

Vocabulary and Phrases

8. **quote** [kwot] *v.* 報價；開價

   How much are they quoting us for the whole package?
   整個方案他們向我們開價多少？

9. **substantial** [səbˋstænʃəl] *adj.* 大量的；多的

   Over the past six months, the company has experienced a substantial increase in orders.
   過去六個月公司增加了大量的訂單。

10. **ramp-up** [ˋræmpˏʌp] *n.* （此指產能或業績的）提升；增加

    If we don't get a ramp-up in sales soon, we might be looking for new jobs.
    如果我們不快點提升業績，我們可能就要找新工作了。

11. **materialize** [məˋtɪriəˏlaɪz] *v.* （願望、計畫）實現

    Hopefully, a better offer will materialize.
    希望可以有比較好的報價。

12. **extrapolate** [ɪkˋstræpəˏlet] *v.* 推斷

    We are using last year's sales numbers to extrapolate over the next two years.
    我們正用去年的銷售數字來推斷未來兩年的情形。

13. **rosy** [ˋrozɪ] *adj.* 樂觀的；美好的

    Dan seems to have a rosy opinion of our prospects.
    丹似乎很看好我們的前景。

14. **valid** [ˋvælɪd] *adj.* 有根據的；有效的

    I still don't know how valid these numbers are.
    我還不知道這些數字有多少根據。

Language Corner

Making the Grade

FETS looks to have a quick ramp-up to healthy business targets.

You Might also Say . . .

- FETS looks to be on course to meeting healthy business objectives.
- FETS seems to have made some improvements with higher business targets in mind.
- FETS appears to be making progress towards its business goals.

Biz Focus

★ commission order 抽佣式的訂單

★ gearmotor 齒輪馬達（亦可寫成 gear motor）

★ retainer fee 聘用費

★ BV 私人有限公司（為荷蘭文 Besloten Vennootschap 的縮寫）

Unit Two
Operations 實際運作

Richard 和總公司對 FETS 未來的營運方向有不同的意見,他希望總公司不要介入子公司日常的業務運作,於是積極和總公司溝通以期建立對內及對外的運作模式,讓參與其中的人能有所依循。

✔ Learning Goals

1. 公司內部必需有完善的稽核制度

2. 公司對外聯絡應由專人負責,避免造成客戶混淆

3. 代理商需與客戶簽訂防止條款,以免客戶直接向工廠下單

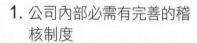
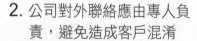

Before We Start 實戰商務寫作句型

Topic 1: Expressing One's Viewpoint 闡述個人觀點

* We appear to have different [conceptions of / ideas of / viewpoints on / points of view on / perspectives on . . .] . . .
  我們對……有不同的 [想法 / 觀點……]。

* I'll try to make sure we can find common ground / come to an agreement / reach a consensus on . . .
  我會試著確保我們在……方面能達成共識。

* I hope you won't mind if I'm candid / direct / frank / straightforward.
  恕我直言。

* Maybe I wasn't clear enough / didn't make myself understood / didn't make my point clear in my communications with (someone), but . . .
  或許在和（某人）溝通時，我的意思沒有表達清楚，但是……

* We can see no [benefit / advantage / use / value . . .] in + V-ing . . .
  我們可以預見……是沒有 [益處 / 效用 / 價值……] 的。

* . . . will simply muddy the waters / cloud the issue / make things more confusing.
  ……只會讓情況更混亂。

Topic 2: Asking for Opinions 尋求意見

* I'll look forward to your thoughts / opinions / ideas on the matter.
  我期待聽到你對此事的想法。

* Please tell me a bit more about / concerning / regarding . . .
  請多告訴我一些有關……的事。

* Is there a particular [arrangement / rule / restriction . . .] that I should be aware of / know of (about) / be informed of?
  有特別的 [協議 / 規定 / 限制……] 是我該知道的嗎？

* I'll send you a [contract draft / quote / proposal . . .] for your comments / feedback / review / reference.
  我會寄一份 [草約 / 報價單 / 計畫書……] 給你，請你提供意見 / 供你參考。

A. Operation Plan 營運計畫

Richard 寫信給總公司的營運部副總裁 Ted，討論子公司未來的運作模式。

September 20

Dear Ted,

Thanks for your comments regarding Far East Technical Services operations. Unfortunately, we appear to have different **conceptions**[1] of what FETS is or should be, and I'll try to make sure we can find **common ground.**[2] I hope you won't mind if I'm **candid.**[3]

I appreciate your offer to help in certain areas of operations, but we envision FETS as a stand-alone **subsidiary**[4] of Maples Corporation, and we oppose any system or procedure that would make us into another MAT, operationally speaking.

Maybe I wasn't clear enough in my communications with Tony, but we assumed that FETS would handle all FETS business, including invoicing, L / C processing, customer and vendor communications, and shipment planning and **expediting.**[5] We can see no benefit in bringing any MCUS personnel into the FETS business. I want to keep a clean business relationship with FETS customers and vendors. Having them deal with two different companies (FETS & MCUS) will simply **muddy the waters.**[6]

I suggested to Tony that the bank send **duplicate**[7] statements to himself and FETS so we can both **keep score.**[8] I've also discussed with Tony the idea of establishing some sort of status-reporting system. We're **amenable**[9] to any reasonable reporting or monitoring system that Tony and you request.

I look forward to your thoughts on the matter.

Sincerely,
Richard

中譯 九月二十號

親愛的泰德：

感謝你對 Far East Technical Services 營運的意見。很遺憾地，我們對 FETS 的定位有不同的想法，我會試著確保我們能達成共識。恕我直言。

我很感激你願意在某些營運方面提供協助，但我們設想 FETS 是一個附屬於 Maples 公司卻獨立運作的子公司。而就營運層面來說，我們也反對任何會讓我們成為另一個 MAT 的體系或做法。

或許在和東尼溝通時，我的意思沒有表達清楚，但是我們認為 FETS 可以自行處理所有業務，包

括開發票、處理信用狀、與客戶和供應商溝通，以及規劃貨運和催貨。我們可以預見讓任何 Maples 美國總公司的員工介入 FETS 的業務是毫無益處的。我想要和 FETS 的客戶和供應商維持單純的商業關係，而讓他們與兩家公司（FETS 和 Maples 美國總公司）接洽只會讓情況更複雜。

我建議過東尼請銀行將帳單副本寄給他本人和 FETS，以讓我們雙方都保有記錄。我也和東尼討論過建立某種現況通報機制的想法。我們願意接受東尼和你所要求的任何合理的報告或監督系統。

我期待聽到你對此事的想法。

祝 商安

理查

Vocabulary and Phrases

1. conception [kən`sɛpʃən] *n.* 想法；概念

My conception is that the company should expand into other countries.
我的想法是公司應該拓展至其他國家。

2. common ground 共識

I'm sure we'll be able to come to common ground during our meeting.
我確信在會議中我們將能達成共識。

3. candid [`kændɪd] *adj.* 坦率的；直言的

I want you to be candid so I can hear your true feelings.
我希望你能坦率一點，這樣我才能聽見你真正的感受。

4. subsidiary [səb`sɪdɪˌɛrɪ] *n.* 子公司；附屬公司

The newspaper was a subsidiary of the broadcasting corporation.
這家報社曾是那家廣播公司的子公司。

5. expediting [`ɛkspɪˌdaɪtɪŋ] *n.* 迅速處理；促進

Phyllis works in the warehouse and is in charge of shipment expediting.
菲利斯在倉庫工作，負責催貨。

6. muddy the waters 使情況更複雜、混亂

Adding a fourth step to the process will just muddy the waters for the user.
在這個程序中增加第四個步驟只會讓使用者覺得更複雜。

7. duplicate [`djupləkɪt] *adj.* 副本的；複製的

I lost my driver's license, so I applied for a duplicate one.
我遺失了我的駕照，所以我又申請了一張。

8. keep score 記錄

I invest in stocks, and I use the Internet to keep score.
我有投資股票並用網路作記錄。

9. amenable [ə`minəbl] / [ə`mɛnəbl] *adj.* 順從的；願意接受的

Her boss was amenable to her request for a raise.
她的老板願意依她的要求加薪。

Language Corner

How to Express a Vision

We envision FETS as a stand-alone subsidiary of Maples Corporation.

You Might also Say . . .

- We foresee FETS as (being) an independent division of Maples Corporation.
- We visualize FETS as (being) a self-sufficient affiliate of Maples Corporation.

B. Business Opportunities 合作機會

離職的老同事寫信給 Richard，表達希望能和他剛成立的公司 FETS 合作。

To: Richard Barlow
From: Eric Lee
Date: December 10
Subject: Future Projects

Dear Richard,

You mentioned that you and Tony had **struck a deal**[1] concerning sourcing and inspecting. Please tell me a bit more about Far East Technical Services and send me, if possible, a few **brochures**[2] that I can selectively **spread**[3] around. How can we work together on future projects? Will customers have to go through Maples US or can we work **independently**?[4] Is there a particular commission arrangement that I should be aware of? This is new to me; however, I believe that there will be lots of opportunities where we can work together.

I haven't sent an initial letter to Maples US HQ because I wanted to ask you these questions without putting you in a particularly difficult position with anyone **on staff**.[5] My intention isn't to **bypass**[6] the system, but to understand it first before we go forward.

Feel free to communicate with me through e-mail at any time. I look forward to hearing from you soon.

Warm regards,
Eric

 收件人：理查 · 巴洛

寄件人：李艾瑞克

日期：十二月十號

主旨：未來計畫

親愛的理查：

你曾提過你和東尼已就採購和品管檢驗的工作達成協議。請多告訴我一些有關 Far East Technical Services 的情形，如果可能的話請寄給我一些宣傳小冊子，讓我可以作選擇性的發放。針對未來的合作案，我們該如何配合？客戶必需透過 Maples 美國總公司嗎？還是我們可以獨立作業？有什麼特別的佣金協議是我該知道的嗎？這對我來說是全新的經驗，但我相信我們將來會有很多合作的機會。

我還沒發信給 Maples 美國總公司，因為我想先詢問你這些問題，以免讓你與公司內部人員間陷入異常困難的處境。我的目的不是要避開這個體制，而是想在我們進一步合作之前先做了解。

隨時都可以用電子郵件與我聯絡。我期待儘快接到你的回信。

謹上
艾瑞克

Vocabulary and Phrases

1. strike a deal 達成協議

The two companies struck a deal to work together.
這兩家公司已達成合作的協議。

2. brochure [bro`ʃur] *n.* 小冊子；宣傳品

The hotel's free brochure convinced the couple to stay there.
飯店免費的宣傳小冊子說服了那對夫婦入住。

3. spread [sprɛd] *v.* 散布；傳播

I'm going to spread the e-mail around the office so people can be informed.
我要將這封電子郵件傳送到整個辦公室，讓大家都收到通知。

4. independently [ˌɪndɪ`pɛndəntlɪ] *adv.*
獨立地；自主地

I'm able to work with a team, but I prefer to work independently.
我可以參與團隊工作，但我偏好獨立作業。

5. on staff 在職的；編制內的

Do we have anyone on staff who knows a lot about computers?
我們公司有哪位員工對電腦很熟悉嗎？

6. bypass [`baɪˌpæs] *v.* 避開；置……於不顧

She bypassed company rules by completing the sale on her own.
她不顧公司規定，自行完成了這次的銷售。

Language Corner

Pursuing Business Opportunities

I believe that there will be lots of opportunities where we can work together.

You Might also Say . . .

- I believe that there will be plenty of chances for us to cooperate / collaborate.
- I think that we can work well in a partnership.
- I'm confident we will make a successful team in the future.

C. Agency Contracts 代理合約

#1 Richard 請美國總公司準備代理採購的合約書。

To: Tony Hughes
From: Richard Barlow
Date: 4/10
Subject: Contractual Discussions between FETS and IBK

Dear Tony,

Before IBK representatives come here, I think we need a contract for them to sign which protects us from having them deal directly with our vendors after we introduce everyone. I think one of the MC legal staff should prepare a contract draft which locks our customers into us as their agent (I don't know whether we should go for a **blanket**[1] **exclusivity**[2] or for specific deals, defined in some way). I understand that the contract should **specify**[3] which country's laws will **prevail**[4] in the event of a **contestation**[5] since the **forum**[6] is international, and I'm told there are **boilerplate**[7] contracts for this sort of thing.

Later,
Richard

中譯

#1

收件人：東尼‧休斯
寄件人：理查‧巴洛
日期：四月十號
主旨：FETS 與 IBK 之間的合約討論

親愛的東尼：

在 IBK 的代表過來之前，我想我們需要準備一份合約讓他們簽以保障我們自己，避免在介紹大家認識之後，他們直接與我們的供應商交易。我想 Maples 美國總公司的法務人員需要準備一份草約，內容約束我們的客戶只能透過我們作為其代理商（我不知道我們是否應該取得全面性獨家代理權，或是在特定交易中另外界定代理權的定義）。既然仲裁法庭是跨國的，我了解合約應該特別指出在有爭議的情況下將採用哪一國的法律，我也被告知有一些這類合約的範本。

再聯絡
理查

Vocabulary and Phrases

1. blanket [ˈblæŋkɪt] *adj.* 適用於所有情況的；沒有限制的
The executive agreed to a blanket salary increase.
經理同意全體加薪。

2. exclusivity [ˌɛksˌkluˈsɪvətɪ] *n.* （此指）獨家代理權
The agent wouldn't do business with the company, unless it agreed to exclusivity.
這家代理商不會和那間公司做生意，除非它同意提供獨家代理權。

3. specify [ˈspɛsəˌfaɪ] *v.* 具體指定；明確說明
He didn't specify what he was upset about.
他沒有明確說明他在生什麼氣。

4. prevail [prɪˈvel] *v.* 佔優勢；盛行
When national and county laws are different, the national law prevails.
當國家法律與郡法有出入時，適用國家法律（國家法律佔優勢）。

5. contestation [ˌkɑnˌtɛsˈteʃən] *n.* 爭議
There is a contestation over the terms of the con
合約的條款有爭議。

6. forum [ˈforəm] *n.* 法院
We're not sure which country's laws will apply the international forum.
我們不確定在國際法庭上將適用哪一個國家的法律。

7. boilerplate [ˈbɔɪlɚˌplet] *n.* 制式範本
There are some free wills on the boilerplate fo to download.
有一些免費的遺囑範本可供下載。

8. draw up 草擬
I'll have my assistant draw up a contract by next v
我會請我的助理在下週前草擬一份合約。

#2 Richard 與客戶（IBK）代表商談代理採購簽約事宜。

To: William Kay
From: Richard Barlow
Date: 4/12
Subject: Contract Update

Dear Mr. Kay,

I've asked our attorneys in the States to **draw up**[8] something which would protect both of us not only in the gearmotor case but for all our future dealings. I sense that you are **wary**,[9] and rightfully so, of an organization such as FETS going directly to your customers. **Conversely**,[10] we have to protect ourselves against the possibility of a customer going directly to our vendors after the introductions are made. I'll send you a draft of the contract for your comments as soon as I receive it.

Regarding the timing of your visit, I'll be tied up from May 6 to May 14, so anytime other than that **stretch**[11] of time is fine with me.

Best regards,
Richard

中譯

#2
收件人：威廉‧凱
寄件人：理查‧巴洛
日期：四月十二號
主旨：合約最新消息

親愛的凱先生：

我已經請我們在美國的律師草擬可以保護我們雙方的合約，這不僅可針對這次齒輪馬達的案子，也適用於我們將來的所有交易。我可以感覺到你對於像 FETS 這類公司直接接觸貴公司客戶的事情很謹慎，這也是理所當然。相反地，我們也必需保護我們自己，避免在互相介紹之後，客戶直接聯繫我們供應商的可能性。我一收到草約就會立刻寄給你，請你提供意見。

有關你來訪的時間，我從五月六號開始到五月十四號將會非常忙碌，那段以外的時間我都有空。

謹上
理查

9. **wary** [ˋwɛrɪ] *adj.* 小心翼翼的；謹防的

I'm always wary of strangers when I walk at night.
當我在晚上走路時，我總會謹防陌生人。

0. **conversely** [kənˋvɝslɪ] *adv.* 相反地

I show them respect; conversely, they respect me as well.
我尊重他們；相反地，他們也尊重我。

1. **stretch** [strɛtʃ] *n.* 持續的一段時間

I'll be out of town for a two-week stretch starting Tuesday.
從星期二開始我要離開這裡兩個禮拜。

Language Corner

1. Talking about a Future Scenario

in the event of + N. / in the event that S. + V. 如果；萬一

For example:

1. We need to be prepared in the event of a typhoon.

= We need to be prepared in the event that a typhoon hits.

2. Saying You'll Be Unavailable

I'll be tied up from May 6 to May 14, so anytime other than that stretch of time is fine with me.

You Might also Say . . .

▪ I'll be unavailable / unreachable from May 6 to May 14, so anytime other than that time period is okay with me / works for me.

Unit Three
Company Profile 公司簡介

新公司要尋找客源、擴展市場，需要準備公司簡介來讓別人了解其所提供的產品與服務。一般公司簡介除了提供給潛在客戶外，也可以供銀行、投資者、媒體或應徵者參考之用。好的公司簡介可以提升公司的專業形象，爭取更多的客戶上門。

✔ Learning Goals

公司簡介的內容通常包含：
1. 歷史沿革及規模大小
2. 產品種類和服務項目
3. 客戶群
4. 經營團隊的背景資料及專長
5. 經營理念與未來展望

Before We Start 實戰商務寫作句型

Topic: Introducing Your Company 介紹公司

The Background:

* Our company was established / founded / started / launched in (year) in (place).
  我們公司於（某年）成立於（某地）。

* We are [pleased / excited / proud . . .] to offer / provide / extend / make available / introduce the following services . . .
  我們 [很高興／驕傲……] 能提供／推出以下服務……

* We maintain close [affiliations / contact / relations . . .] with . . .
  我們與……保持密切 [聯繫／交往／關係……]。

* We have the contacts and [available / accessible / obtainable . . .] resources to accomplish / perform / carry out / execute . . .
  我們有門路與 [可利用／可取得的……] 資源能完成／履行……

The Staff:

* Our company is staffed by / has a staff of . . .
  我們公司有……員工。

* (Someone) graduated with / earned / got his (her) [BSME / BA / MBA . . .] from (school) in (year).
  （某人）在（某年）於（某校）取得他（她）的 [機械工程學士／文學士／企管碩士……] 學位。

* (Someone) has accumulated / gained (number) years of experience as . . .
  （某人）累積（幾）年當……的經驗。

* (Someone) [is in charge of / is responsible for / is accountable for / manages / directs / heads (up) . . .] . . .
  （某人）[掌管／負責／管理／主導／帶領……] ……

The Mission:

* Our goal / aim / ambition / objective is to provide you with . . .
  我們的目標是要提供您……

THE COMPANY

Far EastT Technical Services (FETS) was established in 1996 in Taichung, which is one of the three major industrial areas in Taiwan. Our central location **affords**[1] us easy access to vendors in all parts of the island.

FETS is a technical operation **staffed**[2] by first-rate professional engineers and international business specialists. During the past five years, we have been an OEM supplier* to a steady client base of large corporations, such as:

- Countryside Steel, Ltd. (Holland)
- Allied Computer Systems (US, Canada, and **Australasia**)[3]
- North American Electronics (US and Canada)
- Great Britain Manufacturing (UK)
- United Distributors (US and UK)
- Eastern Metal Suppliers (Japan)

Now, FETS is offering our services to companies large and small with the aim of establishing efficient, mutually beneficial relationships with a broad range of business partners. We are pleased to offer the following services:

- **Component**,[4] assembly, and product sourcing
- Quality-control inspections
- Vendor surveys
- Plastic injection molding
- Engineering and QC consultation
- Production management
- Patent service
- Shipping paperwork assistance

In addition to our **in-house**[5] technical expertise, FETS maintains close **affiliations**[6] with several other engineering service organizations in Asia, so we have the contacts and available resources to accomplish **virtually**[7] any assignment. Also, each of the FETS staff reads, writes, and speaks fluent English. Quotes, test reports, and daily communication are always clear, concise, and correct.

 Far East Technical Services 簡介

公司沿革

Far East Technical Services (FETS) 於 1996 年成立於台灣三大主要工業區之一的台中。處於核心位置使我們便於與全台各地的廠商接觸。

FETS 是一家擁有優秀專業工程師與國際貿易專家的技術服務公司。過去五年來，我們成為一間具有穩定客群的代工生產商，我們的客戶包含下列大型企業：

- Countryside Steel, Ltd.（荷蘭）
- Allied Computer Systems（美國、加拿大與澳大拉西亞）
- North American Electronics（美國與加拿大）
- Great Britain Manufacturing（英國）
- United Distributors（美國與英國）
- Eastern Metal Suppliers（日本）

現在 FETS 提供服務予大小型公司，以與廣泛的商業夥伴建立有效率、互利的關係為目標。我們很高興能提供以下服務：

- 零組件和產品的採購
- 品管檢驗
- 訪查供應商
- 塑膠射出成型
- 技術和品管諮詢
- 生產管理
- 專利權服務
- 幫忙處理出貨文件

除了我們公司內部的專業技術之外，FETS 也和亞洲其他一些技術服務公司保持密切聯繫，因此我們有門路和可運用的資源來完成幾乎所有的任務。再者，每一位 FETS 員工的英文讀寫及口說能力都十分流利。報價、測試報告和日常溝通向來都很清楚、簡潔且正確。

▶ Vocabulary and Phrases

1. **afford** [əˋfɔrd] *v.* 提供；給予
   Owning my own company affords me a lot of freedom.
   擁有自己的公司給我很大的自由。

2. **staff** [stæf] *v.* 擔任……的職員；為……配備職員
   Our store is staffed by six salespeople.
   我們店裡有六名銷售員。

3. **Australasia** [ˌɔstrəˋleʒə] *n.* 澳大拉西亞（泛指澳紐及附近南太平洋諸島）
   It's common to see companies in Australasia and China working together.
   澳大拉西亞和大陸兩區的公司合作是很普遍的。

4. **component** [kəmˋponənt] *n.* 零件；成分
   She bought stereo components from three different stores.
   她在三間不同的商店買了立體音響的零件。

5. **in-house** [ˋɪnˌhaʊs] *adj.* （公司、組織）內部的
   We have an in-house customer service staff.
   我們公司內部有客服人員。

6. **affiliation** [əˌfɪliˋeʃən] *n.* 聯繫；加入
   We have no affiliation with that company.
   我們和那間公司沒有聯繫。

7. **virtually** [ˋvɝtʃʊəlɪ] *adv.* 幾乎；事實上
   He owns virtually all the land in the area.
   他幾乎擁有那區所有的土地。

▶ Language Corner

Expressing the Goal

with the { aim / objective / goal / intention } of doing sth
with an eye to (doing) sth
以……為目的、目標

For example:

1. The company increased funding with the aim of improving profits.
2. With an eye to helping the environment, Kathy recycled as much as possible.

▶ Biz Focus

★ **OEM supplier** 提供代工生產的供應商（OEM 為 original equipment manufacturer 的縮寫）

THE MANAGEMENT STAFF

RICHARD BARLOW, our managing director, began his engineering career in 1983 with Lette Company as a test engineer. In 1991, he joined Twin Brothers as a senior test engineer, and shortly afterward was promoted to Manager of Reliability Engineering. Then, as Director of Far East Operations for Maples Corporation, Richard received the **exposure**[8] to the Far East that enabled him to come to Taiwan to establish what is now Far East Technical Services.

JACK HUANG, who is in charge of our sourcing department, earned his BSME from National Taiwan University in 1992. After spending two years working for Philips NV* as an engineer, Jack was awarded an opportunity, through an island-wide testing and selection process, to represent Taiwan as part of an international collection of engineers on a steel-making project in Indonesia. In 1997, the project ended and Jack returned to Taiwan and joined FETS, where he has since gained excellent experience in sourcing, tooling, packaging, and manufacturing.

HARRY TSAI, who manages our quality-control inspectors in Taiwan and China, has an engineering degree and five years of experience as a QC engineer with Anytech Corporation, a communications equipment manufacturer. Harry spent three of his five years with Anytech in Australia, where he directed the QC department at its Melbourne plant. Harry returned to Taiwan in 1999 and started to work for FETS.

MONICA YANG, our office manager since 1999, earned a BA in business administration from Taichung Commercial College in 1989 and a BA in English, with a computer science minor, from Providence University in 1994. Monica has **accumulated**[9] nine years of experience with trading and manufacturing companies as a business secretary and administrative manager.

THE PHILOSOPHY

We understand the challenges of doing business from an Asian country because we have extensive experience ourselves. And we realize that the risks are substantial without your own on-site **presence**.[10] Our goal is to provide you with a partnership in Asia that most smaller companies cannot afford to establish.

We're small enough to be able to give you **personalized**[11] service. We're big enough to provide stable, comprehensive service now and into the future.

So whether your interest is in finding components at a lower cost, manufacturing a whole product, improving quality, buying tooling, or monitoring your vendors, Far East Technical Services can help to get your dreams on the production line.

Call us at any time if you're interested in the services we offer.

中譯

管理階層

理查・巴洛為我們的總經理,他於 1983 年在 Lette 公司擔任測試工程師,開始其工程師生涯。1991 年他加入 Twin Brothers 擔任資深測試工程師,不久之後便晉升為可靠度工程經理。之後,在擔任 Maples 公司遠東區營運總監時,理查有機會接觸到遠東區業務,並來到台灣建立 Far East Technical Services 這家公司。

黃傑克掌管我們的採購部門,他在 1992 年於台灣大學取得機械工程學士。在荷蘭 Philips 上市公司擔任兩年的工程師之後,透過全國性的考試與甄選程序,傑克獲得機會代表台灣與一群國際工程師一同參與在印尼的煉鋼計畫。1997 年該計畫結束後,傑克回到台灣並加入 FETS,從此累積了豐富的採購、模具、包裝和製造的經驗。

蔡哈利負責管理我們在台灣和大陸的品管檢測員。他擁有工程學學位和在 Anytech 公司(一間通信設備製造商)五年的品管工程師經驗,其中三年他帶領 Anytech 在澳洲墨爾本廠的品管部門。哈利在 1999 年回到台灣並開始為 FETS 工作。

楊莫妮卡從 1999 年以來便擔任我們的辦公室經理。她在 1989 年於台中商專取得企管學士學位;1994 年於靜宜大學取得英語學士學位,並副修電腦科學。莫妮卡在貿易和製造公司累積九年擔任商務秘書和行政經理的經驗。

宗旨

我們了解在亞洲國家做生意的挑戰,因為我們有很豐富的經驗。而我們也知道沒有自己的人員派駐在工廠有很大的風險。我們的目標是和您建立在亞洲的合夥關係,而這是大多數較小型的公司無法做到的。

我們的規模小到可提供符合您個別需求的服務;規模大到能在現在與未來提供穩定、廣泛的服務。

因此無論您的興趣是尋找較低成本的零件、製造完整產品、改善品質、採購模具或監督您的供應商,Far East Technical Services 都可以幫得上忙。

如果您對於我們提供的服務有興趣的話,敬請隨時與我們聯絡。

▶ Vocabulary and Phrases

8. **exposure** [ɪk`spoʒɚ] *n.* 接觸;暴露

   As an intern, he received exposure to our sales methods.
   當實習生時,他接觸到我們的銷售方法。

9. **accumulate** [ə`kjumjəˌlet] *v.* 累積

   He accumulated 10 hours of overtime this week.
   他這星期累積了十小時的加班時數。

10. **presence** [`prɛzns] *n.* 到場;存在

    The company would like to establish a presence in the Middle East.
    這間公司想要打進中東市場。

11. **personalized** [`pɝsnəlˌaɪzd] *adj.*
    量身訂做的;符合個人需求的

    All customers will receive personalized e-mails thanking them for their purchase.
    所有顧客都會收到個人化的電子郵件以感謝他們的購買。

▶ Biz Focus

★ NV 股票有公開發行的股份有限公司(為荷蘭文 Naamloze Vennootschap 的縮寫)

Unit Four
Networking 擴展商機

Richard Barlow 為了爭取代理 Senture Products Company 在台採購及品管檢驗的工作,特別寫了這篇說帖,詳述 FETS 可提供的服務項目,並分析其中的利弊得失,以讓客戶了解委託 FETS 代為處理台灣的業務,可以省掉許多麻煩和費用。

 Learning Goals

1. 外商企業在異國經商可能會面臨的種種困難

2. 外國人在異地居住可能會有的食、衣、住、行及文化衝擊等問題

3. 提供品管檢驗服務時,需依人力多寡、檢驗設備及使用空間大小,來核算應收取的費用

Before We Start 實戰商務寫作句型

Topic 1: Explaining Benefits 說明好處

* . . ., with no [capital investment / start-up funds / additional payment . . .] required / needed / necessary.
  ……，不需要 [資本投資 / 創業基金 / 額外費用……]。

* There will be no concerns / trouble / hassles / complications about [visas / housing / transportation . . .].
  沒有 [簽證 / 住所 / 交通……] 方面的擔憂 / 麻煩 / 困難。

* You will benefit (profit) from / have the benefit (advantage) of . . .
  你將從……獲益。

* Your company will have the capability / capacity / ability to V. . . .
  貴公司將有能力……

Topic 2: Explaining Fees 解釋費用

* We propose to supply the following services for a [monthly / yearly (annual) / biannual . . .] fee of (amount of money) . . .
  我們打算以 [每月 / 年 / 半年……]（金額）的費用來提供以下服務……

* The [cost / expense / charge / price . . .] of . . . is not included in / excluded from / omitted from these proposed fees.
  ……的 [費用 / 價錢……] 不包含在這些提出的費用中。

* Fees [presented / listed / specified . . .] in this proposal contain / cover . . .
  本計畫書所 [提出 / 列舉 / 詳述……] 的費用包含……

* Expenses for . . . will be billed / charged / invoiced separately / discretely.
  ……的費用會個別計費 / 開發票。

* A charge / fee of (amount of money) will be required to cover expenses for / pay for . . .
  將需要（金額）的費用來支付……

The Partnership 合夥關係

November 18

Mr. Ken Green
Senture Products Company
3256 Springfield Ave, Suite 220
Seattle, WA 98765

Dear Ken,

Thank you again for lunch, and the opportunity to chat a bit about our operations here in Taiwan. I've attached an overview (Attachment #1) **highlighting**[1] the benefits of hiring FETS as well as the **downside**[2] of employing a direct-hire engineer and a proposal (Attachment #2) covering the matters we discussed.

As I mentioned to you during our meeting, I think it's clear upon review of the facts that hiring FETS to represent you **beats**[3] directly hiring a person to live in Taiwan.

I think we have the ability to form a successful partnership. Thank you for your consideration.

Sincerely,
R.A. Barlow
R.A. Barlow

中譯

十一月十八號

肯・葛林先生
Senture Products 公司
98765 華盛頓州西雅圖市
春田大道 3256 號 220 室

親愛的肯：

再次感謝您能一起共進午餐和給我們機會談論我們在台灣營運的情況。我附上一份概要（附件一）— 強調雇用 FETS 公司的優點與直接聘請工程師的缺點，和一份計畫書（附件二）— 涵蓋了我們之前討論的內容。

如我在會面時所提到的；綜觀各方面的事實，我想我們都清楚聘用 FETS 代表貴公司勝過直接在台灣派駐一名員工。

我認為我們可以建立一個成功的夥伴關係。謝謝您的考慮。

祝 商安
R.A. 巴洛

Vocabulary and Phrases

1. highlight [ˈhaɪˌlaɪt] *v.* 強調；使突出

During the meeting, the CEO highlighted the achievements of some employees.
執行長在會議中強調了一些員工的成就。

2. downside [ˈdaʊnˌsaɪd] *n.* 不利的一面

The downside of living downtown is that it's expensive.
住在市中心的缺點是費用太高。

3. beat [bit] *v.* 勝過；打敗

Taking the subway beats sitting in traffic while riding the bus.
搭地鐵勝過在塞車時搭公車。

Attachment #1

Overview

BENEFITS OF HIRING FAR EAST TECHNICAL SERVICES INSTEAD OF A DIRECT-HIRE ENGINEER

1. Instant start-up of a **turnkey**[4] operation, with no capital investment required.

2. No need to obtain government business licenses, rent spaces or buy equipment for your office or lab.

3. No concerns about visas, housing, transportation, culture shock, or cost of living.

4. No **hassles**[5] related to employee **recruitment**,[6] hiring, supervision, insurance, or income taxes, which you would all need to deal with if you hired a local. Also, your FETS service will be uninterrupted regardless of employee **turnover**.[7]

5. The benefit of having a Chinese engineer working with Chinese vendors in his / her native language, closely supervised by senior Chinese and American engineers and managers.

6. Capability to expand overseas operations easily and quickly to reflect the growth of Senture.

7. Excellent communication, written and spoken in English, with the Senture office.

OTHER FACTORS TO BE CONSIDERED

Living in Taiwan is a very expensive **proposition**,[8] especially for an American who wants Western-style food and **lodging**.[9] Cost of living is a key factor to consider when evaluating the expense of keeping an expatriate in a foreign country, and the quality of life, which is related to the cost of living, is also very important. Making sure an American is happy enough to want to live in a decidedly different environment isn't easy.

Taipei was recently named the eighth most expensive city in the world in which to live (Manhattan is #34). It costs an employer US$9,000 to US$11,000 per year to put an employee's child in an American primary or high school here. Cars cost twice as much as they do in the States. A three-bedroom apartment **comparable**[10] to those in a big American city is two to three times more expensive in Taiwan. Prices of Western food and drink are three to seven times higher than in the States. The price of doing business in Taiwan, considering office rents and equipment expenses, is **astronomical**.[11]

 中譯 附件一

概要

雇用 FETS 公司而非直接聘請一名工程師的好處

1. 勿需投入資金,可立即開始運作。

2. 不必取得政府核發的營業執照、租賃辦公室或實驗室場所或購買設備。

3. 沒有簽證、房屋、交通、文化衝擊或生活開銷等擔憂。

4. 可以避免雇用當地員工所要處理如員工招募、雇用、管理、保險或所得稅等麻煩。再者,FETS 公司的服務是不會因員工流動而中斷的。

5. 可享有華人工程師直接用母語和華人供應商工作的好處,並受資深華人及美國工程師和經理的嚴密監督。

6. 可輕易且迅速地拓展海外的營運以反映 Senture 公司的成長。

7. 能以英語與 Senture 辦公室作良好的書面及口語溝通。

其他考慮因素

居住在台灣有物價高昂的問題,特別是對一位想要西式食物和住所的美國人而言。當估算外派人員到國外的費用時,生活開銷是一個要考量的關鍵因素,而與生活開銷相關的生活品質也是非常重要的。要確保一位美國人可以愉快地生活在一個截然不同的環境是不容易的。

台北最近在全球居住城市的生活開銷排名中名列第八(曼哈頓排名三十四)。一位美國雇主一年需支付九千到一萬一千元美金,才能讓一名員工的小孩就讀這裡的美國小學或中學。台灣汽車的價錢為美國的兩倍。可和美國大城市相比的三房公寓,在台灣要貴上兩到三倍。西式飲食的價格比美國高出三到七倍。考量到辦公室租金和設備費用,在台灣做生意的開銷是相當龐大的。

Vocabulary and Phrases

4. turnkey [ˈtɝnˌki] *adj.*

可立即開始工作的;組裝完備的

We have a turnkey group of computer programmers available to us at any time.

我們隨時都有一組電腦程式設計師在待命。

5. hassle [ˈhæsl̩] *n.* 麻煩;困難

It's such a hassle to go through customs at the airport.

在機場通關真是一件麻煩的事。

6. recruitment [rɪˈkrutmənt] *n.* 招募新人

The recruitment of new workers is not going as quickly as we expected.

招募新人的進度不如我們預期的快速。

7. turnover [ˈtɝnˌovɚ] *n.* 員工流動率

The company has high turnover because they treat their workers poorly.

這間公司的員工流動率很高,因為他們對員工很不好。

8. proposition [ˌprɑpəˈzɪʃən] *n.*

(要處理的)事情;問題

Trying to convince Ken he's wrong is a difficult proposition.

要說服肯他是錯的是一件很困難的事。

9. lodging [ˈlɑdʒɪŋ] *n.* 住所

She's new in town and is still looking for permanent lodging.

她剛來到這裡,還在找長期居住的地方。

10. comparable [ˈkɑmpərəbl̩] *adj.*

可相比的;比得上的

Our chef is comparable to anyone working in a five-star hotel.

我們的主廚可媲美任何在五星級飯店工作的廚師。

11. astronomical [ˌæstrəˈnɑmɪkl̩] *adj.*

(數量、金額)龐大的

The price of fuel in that country is astronomical.

那個國家的燃料費很高。

PROPOSAL

Far East Technical Services proposes to supply the following services to Senture Products Company for a monthly fee of US$5,000:

1. Assign a full-time, experienced engineer to work with Senture's vendors in Taiwan and China with the goal of improving zero-hour quality* of Senture's products.

2. Equip the Senture representative with appropriate technical, **managerial**,[12] and administrative support.

3. Provide 250 sq. ft. of lab space for Senture's reliability engineering tests (additional space available).

4. Supply up to two **man-hours**[13] per day of technician time for incidental lab testing.

5. Furnish written reports at reasonable time **intervals**[14] as required by Senture.

RESTRICTIONS AND NOTATIONS:

1. The FETS laboratory is equipped with usual test equipment, such as chart recorders,* force- and temperature-measuring instruments, and common hand tools. The cost of special-purpose test equipment is not included in these proposed fees.

2. Fees presented in this proposal include employee benefits and on-island travel and incidental expenses. Expenses for travel to China will be billed to Senture separately.

3. FETS proposes to provide services on a monthly contract basis, payable in advance each month. We also propose that the contract be cancelable by either party with one month's notice.

4. A set-up charge of US$3,500 will be required to cover expenses for FETS management to survey each of Senture's vendors (approx. 12) in Taiwan and China.

中譯 附件二

計畫書

Far East Technical Services 打算以每月五千元美金的費用來提供 Senture Products 公司以下的服務：

1. 指派一名全職、有經驗的工程師與 Senture 在台灣和大陸的供應商合作，以期改進 Senture 產品的零小時品質（出貨檢驗時的品質）。

2. 給予 Senture 代表人員適當的技術、管理和行政支援。

3. 提供 250 平方英尺的實驗室空間作為 Senture 可靠度工程測試之用（需額外的空間亦可提供）。

4. 每天提供技術人員多達兩人工時以處理偶發的實驗室測試。

5. 依 Senture 公司的需求，在合理的時間內提交書面報告。

限制與註記：

1. FETS 公司的實驗室備有一般測試設備，如資料自動記錄器、測量力量與溫度的儀器及常見的手動工具。特殊用途的檢驗設備之花費不包含在這些提出的費用中。

2. 本計畫書所列的費用包括員工福利、島內交通和雜費。去大陸的差旅費會向 Senture 公司另外收費。

3. FETS 公司打算以按月計費的合約為基礎來提供服務，每個月事先支付。我們也建議任一方在一個月前事先告知對方的情況下可取消合約。

4. 需收三千五百元美金的開辦費來支付 FETS 公司管理階層訪查 Senture 在台灣和大陸的供應商（大約十二家）的費用。

Vocabulary and Phrases

12. managerial [ˌmænəˈdʒɪrɪəl] *adj.*
管理方面的

She's hoping to apply for a managerial position.
她希望應徵管理職。

13. man-hour [ˈmænˈaur] *n.*
一人一小時的工作量；人工作時

The building of the tunnel took hundreds of thousands of man-hours.
興建那條隧道耗費數以十萬計的人工時。

14. interval [ˈɪntəvl̩] *n.* 間隔；距離

I usually have six-month intervals between visits to the dentist.
我通常半年去看一次牙醫。

Biz Focus

★ **zero-hour quality** 零小時品質

產品的品質可分為「現在品質」與「未來品質」。「現在品質」即「零小時品質」，是指產品在製造之後、使用之前的品質狀況，也就是「出貨檢驗時的品質」；「未來品質」指「可靠度」，是指產品在消費者實際使用時，整體所呈現出來的品質。

★ **chart recorder** 資料自動記錄器

Language Corner

Giving a Person Something

provide sb with sth / provide sth for sb
supply sb with sth / supply sth to sb
equip / furnish / outfit sb with sth

提供某人某物

For example:

1. She supplied him with data about her company's products.

2. The government outfitted the army with high-tech trucks.

Chapter Three

Standard Kinds of Letters

Unit One
Congratulations 祝賀

恭賀信函通常都很精簡，開頭
便直接恭賀對方，讚揚、肯定
對方的成就及作為。

 Learning Goals

學習撰寫各種不同的恭
賀信函，如祝賀結婚、
生小孩、業績提升、過
新年等

Before We Start 實戰商務寫作句型

Topic 1: Cheering Them On 鼓勵對方

◆ Wishing you all the best / the best of luck!
祝你事事順利 / 好運！

◆ Our warmest / heartiest / sincerest congratulations on your [10th anniversary / promotion / award / accomplishment . . .].
我們衷心 / 誠摯地恭喜你的 [十週年紀念 / 升官 / 得獎 / 成就……]。

◆ We wish you [continued growth / every happiness in your new home / a happy new year / safe passage . . .].
我們祝你 [持續成長 / 新居落成愉快 / 新年快樂 / 旅途平安……]。

◆ May the New Year bring you [happiness / serenity . . .].
願新年為你帶來 [快樂 / 平靜……]。

Topic 2: Marriage Congratulations 結婚祝賀

◆ Congrats on your engagement!
恭喜你訂婚！

◆ We wish you and your bride-to-be (husband-to-be) a lifetime of happiness!
我們祝你和你的未婚妻（夫）幸福一輩子！

◆ From all of us here at (company), we wish you two a long and happy life together!
（某公司）全體同仁祝你們倆永遠幸福！

◆ Congratulations & best wishes for a very blessed marriage and happy life together!
恭喜並祝福你們婚姻生活非常幸福美滿！

◆ Congratulations! You make a wonderful couple. I wish you happiness for the future.
恭喜！你們真是天作之合。我祝福你們未來幸福愉快。

A. Engagement 訂婚

To: Matt
From: Jonathan
Date: June 20
Subject: **Tying the Knot**[2]

Dear Matt,

    Congrats[3] on the big announcement! We're all happy to hear of your engagement! Have you chosen a date yet for the upcoming **nuptials**?[4] They say that springtime is the best to get hitched. If you need some help with the preparations, I can give you the name of the wedding planner Mary and I used. It's better than putting all that **pressure**[5] of planning on you and your **bride-to-be**.[6] Either way, I'm sure it will be a memorable occasion.

    From all of us here at Maples Corporation, we wish you two a long and happy life together!

Best Wishes,
Jonathan

中譯

收件人：麥特
寄件人：強納森
日期：六月二十號
主旨：締結連理

親愛的麥特：

    恭喜你，宣布了這個大好消息！我們都很高興聽到你訂婚了！即將來臨的婚禮你選定好日期了嗎？他們說春天是最適合結婚的季節了。如果你需要人幫忙準備的話，我可以介紹我和瑪莉的婚禮企劃給你。這會比將所有策劃的壓力都放在你和你未婚妻的身上來得好。無論如何，我相信這都將是一場永生難忘的盛宴。

    Maples 公司全體同仁祝你們倆永遠幸福快樂！

謹上
強納森

Vocabulary and Phrases

1. engagement [ɪnˋgedʒmənt] n. 訂婚；婚約

After an engagement of six months, Rhonda and Bart were married.
蓉妲和巴特在訂婚半年後就結婚了。

2. tie the knot 結婚（= get hitched）

When are you two going to tie the knot?
你們倆什麼時候要結婚？

3. congrats [kənˋgræts] n.（口語）祝賀；恭喜（= congratulations）

Congrats on your retirement, Mr. James.
詹姆士先生，恭喜您退休。

4. nuptials [ˋnʌpʃəlz] n. 婚禮

Nearly 500 people attended Douglas Terry's nuptials.
將近五百人出席道格·泰瑞的婚禮。

5. pressure [ˋprɛʃɚ] n. 壓力

You're putting too much pressure on your son to do well in school.
你要求你兒子在學校有好的表現，給他太多壓力了。

6. bride-to-be [ˋbraɪdtəbɪ] n. 未婚妻

According to the newspapers, the bride-to-be is from China.
根據報紙報導，那位未婚妻來自中國。

B. Newborn Baby 喜獲麟兒

To: Don
From: Sherry
Date: 5/10
Subject: **Addition**[7] to the Family

Dear Don,

I just heard the big news. Congratulations on your newborn baby. I was so excited to learn about the new addition to your family. Have you **come up with**[8] a name for him yet? The first years are the most **precious**.[9] The next time you come to our office, be sure to bring some photos along. I'm sure you must have taken a lot. I'm looking forward to seeing your new **bundle of joy**![10]

I hope both the mother and baby are doing well.

Wishing you all the best!

Yours truly,
Sherry

中譯

收件人：唐
寄件人：雪莉
日期：五月十號
主旨：家庭新成員

親愛的唐：

我剛聽到這個大好消息。恭喜你喜獲麟兒。聽到你們家有新成員加入真是太令人興奮了。你幫他取名字了嗎？頭幾年是最寶貴的。下一次你來我們辦公室時，要記得帶一些照片過來，我確信你一定拍了很多。期待能見到你們的小寶貝！

希望母子均安。

祝你們事事順利！

謹啟
雪莉

---

7. **addition** [əˋdɪʃən] *n.* 增加的人（物）

   She will make a nice addition to the company.
   她的加入對公司來說會是一大加分。

8. **come up with** 想出

   How did you come up with the answer?
   你怎麼想出答案的？

9. **precious** [ˋprɛʃəs] *adj.* 寶貴的；珍貴的

   Since this watch was a gift from my wife, it is very precious to me.
   因為這隻錶是我太太送的禮物，所以對我來說它非常珍貴。

10. **bundle of joy** 新生兒

    What a lovely bundle of joy. How old is he?
    好可愛的寶寶。他多大了？

Language Corner

Giving Congratulations

We're (all / so / really) happy / delighted / excited . . .

For example:

1. We're all happy to hear of your impending nuptials.
2. We're so delighted to learn that you got the job of your dreams.
3. We're really excited about the birth of your son.

To: Ben
From: The MAT Pack
Date: October 2nd
Subject: 10 Years of Success

Dear Ben,

From your friends and admirers in Taiwan, our warmest congratulations on your 10 years of successfully leading Maples Corporation. This **achievement**[2] speaks of the hard work and professionalism that have always been **embodied**[3] by your branch. You have set an impressive standard, one we all hope to **emulate**[4] in the future. You have also been a good friend to so many of us here. It has truly been a pleasure knowing you.

Here's to 10 more years — years that are easier and more **profitable**.[5] We hope that you find them as **rewarding**[6] as ever, and that you are **at the helm**[7] of Maples Corporation for many years to come.

Sincerely,
The MAT Pack

. . . .

To: The MAT Pack
From: Ben
Date: October 2nd
Subject: Re: 10 Years of Success

Dear All,

Thank you very much for celebrating our 10th anniversary with us. We feel honored that you can share in the excitement of this special occasion.

We have gained **invaluable**[8] knowledge from branches like yours and are **indebted**[9] to all you have done. Occasions like this are a celebration of the teamwork that has made us all so successful. We all feel proud to be part of this great team and look forward to further cooperation. These have been exciting times for all of us. I know I speak for everyone on this when I say I appreciate your help and kind thoughts.

Best regards,
Ben

中譯

收件人：班
寄件人：MAT 全體同仁
日期：十月二號
主旨：十年有成

親愛的班：

我們這群在台灣的朋友和仰慕者衷心地恭喜你成功地領導 Maples 公司邁入十週年。這項成就說明了貴單位向來所體現的努力工作與專業精神。你樹立了一個令人欽佩的典範，這是我們希望將來可以效法的。你同時也是我們這裡很多人的好朋友。真的十分高興能夠認識你。

祝你們往後的十年可以更順利、更賺錢，希望接下來的日子對你來說也同樣很值得，同時你也能繼續帶領 Maples 公司。

祝 商安
MAT 全體同仁

. . . .

收件人：MAT 全體同仁
寄件人：班
日期：十月二號
主旨：回覆：十年有成

親愛的各位：

非常感謝你們和我們一起慶祝我們的十週年紀念。我們很榮幸你們可以與我們一同分享這個特別時刻的喜悅。

我們從很多像你們一樣的部門得到了無價的知識，也很感激你們為我們所做的一切。這樣的盛事正是為了慶祝使我們如此成功的團隊合作。我們很驕傲能成為這傑出團隊的一分子，並期待能進一步合作。對我們所有人而言，這真是令人興奮的時刻。我們一致感謝你們的協助和體貼。

謹上
班

Vocabulary and Phrases

1. **anniversary** [ˌænəˈvɝsərɪ] *n.* 週年紀念（日）

   My parents just celebrated their 43ʳᵈ anniversary.

   我父母剛慶祝完他們四十三週年的結婚紀念日。

2. **achievement** [əˈtʃivmənt] *n.* 成就；達成

   Winning a Nobel Prize is an incredible achievement.

   贏得諾貝爾獎是一項很了不起的成就。

3. **embody** [ɪmˈbɑdɪ] *v.* 體現；使具體化

   He doesn't just talk about leadership. He embodies it.

   他不是只談論如何領導。他實際去體現它。

4. **emulate** [ˈɛmjəˌlet] *v.* 效法；努力追趕

   Many companies have tried to emulate Google's business model.

   許多公司都試著效法 Google 的經營模式。

5. **profitable** [ˈprɑfɪtəbl] *adj.* 獲利的；有用的

   Even though they worked hard, their restaurant still wasn't profitable.

   即使他們辛勤工作，他們的餐廳還是沒有獲利。

6. **rewarding** [rɪˈwɔrdɪŋ] *adj.* 有益的；有報酬的

   Doing well in school can be quite rewarding.

   在學校表現良好將會帶來很多好處。

7. **at the helm** 領導；掌控

   With the CEO at the helm, the company has once again become a major force in the computer industry.

   有那位執行長的領導，該公司再次成為電腦產業的龍頭。

8. **invaluable** [ɪnˈvæljəbl] *adj.* 無價的；非常貴重的

   The Internet is an invaluable source of information.

   網路是非常有用的資訊來源。

9. **indebted** [ɪnˈdɛtɪd] *adj.* 感激的；受惠的

   The company is indebted to your service.

   公司很感激你的服務。

Language Corner

Recognizing Peers

You have set an impressive standard, one we all hope to emulate in the future.

You Might also Say . . .

- You have really set a shining example. We hope to emulate your contributions someday.
- You have done something unprecedented. We all hope to achieve a similar level of success in the future.
- You have achieved a breakthrough in your field. We aspire to follow in your footsteps.

D. Sales Growth 業績成長

To: Sales Team
From: Richard Barlow
Date: 9/9
Subject: Sales Targets Met

Dear All,

My heartiest congratulations on the work you have done. The sales numbers you put up last quarter would exceed anyone's wildest expectations. You beat our sales **projections**[1] by a **whopping**[2] eight and a half percent. Once more, you did this while the market was in a **downturn**.[3] You did this even though your expense accounts had been **slashed**[4] **across the board**.[5]

Your achievements have not gone unnoticed. To express our thanks, each member will be receiving NT$30,000 in travel coupons.

Keep up the good work.

Warm regards,
Richard

收件人：業務部
寄件人：理查·巴洛
日期：九月九號
主旨：達成業績目標

親愛的各位：

我衷心地恭喜大家這次的工作表現。上一季你們達到的銷售數字遠遠超過任何人的預期。你們超越了我們的業績預估有百分之八點五之多。你們再一次地在不景氣的時刻，甚至是在你們的經費被全面性大幅刪減的情況下，達到了業績目標。

你們的成就是有目共睹的。為了表達我們的感謝，每位同仁將可獲得價值新台幣三萬元的旅遊優待券。

繼續保持良好的工作績效。

謹上
理查

Vocabulary and Phrases

1. **projection** [prəˋdʒɛkʃən] *n.* 估計；預測

   Without these totals, we can't make a sales projection.
   沒有這些總數，我們無法估計銷售量。

2. **whopping** [ˋhwɑpɪŋ] *adj.* （口語）龐大的；極大的

   We made a whopping NT$23 million on the stock market!
   我們在股市中大賺了新台幣二千三百萬元！

3. **downturn** [ˋdaʊnˏtɜn] *n.* （經濟）衰退；下降

   The reason we're losing money is because we're in the middle of an economic downturn.
   我們賠錢的原因在於我們正處於經濟不景氣。

4. **slash** [slæʃ] *v.* 大幅度刪減

   The store was slashing prices on most of its skin products.
   這間店大部分的護膚產品都在大減價。

5. **across the board** 全面性地

   The new company policy will affect the employees across the board.
   這項新的公司政策將影響所有員工。

6. **cruise** [kruz] *n.* （坐船）旅行；巡航

   Ever since Virginia was a little girl, she has dreamed of taking a cruise around the world.
   維吉尼亞從小就夢想坐船環遊世界。

E. New Year 新年

To: Jan Meyer
From: Daniel Parker
Date: 1/6
Subject: Happy New Year

Dear Mr. Meyer,

I hope everything goes well for you in the upcoming year. I hear you're taking a couple of weeks off to go on a **cruise**.[6] So, I wish you smooth sailing on your vacation and for the year that follows.

This past year was exceptionally successful for all of us here. Hopefully, this year will be even better. We are all counting on the continued cooperation between our companies. May the Chinese New Year, only a month away, also bring you happiness, **serenity**,[7] and success.

All the best,
Daniel Parker
Assistant to the CEO

中譯

收件人：強・梅耶
寄件人：丹尼爾・帕克
日期：一月六號
主旨：新年快樂

親愛的梅耶先生：

希望您來年諸事順心如意。我聽說您要休假幾週搭船去旅行，我祝您旅途平安、新年順利。

過去一年對我們來說是相當成功的，但願今年可以更好。期望我們公司之間能持續合作。願一個月後的中國新年也為您帶來快樂、平靜和成功。

祝 一切安好
丹尼爾・帕克
執行長助理

7. **serenity** [səˈrɛnətɪ] *n.* 平靜
When I go fishing, I enjoy the serenity most.
當我去釣魚時，我最享受那平靜的時刻。

Language Corner

1. Great Results

The sales numbers you put up last quarter would exceed anyone's wildest expectations.

You Might also Say . . .

- The sales figures you posted last quarter were great by anyone's standards.
- The sales results you had last quarter were way beyond anyone's expectations.

2. High Hopes

I hope everything goes well for you in the upcoming year.

You Might also Say . . .

- I hope everything goes according to plan in the coming year.
- I expect everything will work out for you in the new year.

Unit Two
Appreciation 感謝

私人謝函可以手寫,但代表公司的謝函則需打字書寫在有公司抬頭的信紙上。謝函通常會指名寫給特定人士,內容可以精簡,但需充分讓對方感受到你誠摯的謝意。

✔ Learning Goals

學習撰寫感謝信函

Before We Start 實戰商務寫作句型

Topic: Showing Thanks 表示感謝

- We are very grateful / thankful / obliged / indebted to you for your . . .

- Thank you so much / Many thanks for + V-ing . . .
  （我們）非常感激你……

- I would like to take this opportunity / chance to express my most heartfelt / sincerest / most earnest appreciation to you for your [hard work / assistance / hospitality / support . . .].
  我要藉此機會來表達我對你的 [努力工作 / 協助 / 熱情款待 / 支持……] 最誠摯的謝意。

- We feel very fortunate / lucky to V. . . .
  我們覺得非常幸運能夠……

- Owing to / Due to / Thanks to your [thoughtfulness / kindness / helpfulness / unselfishness . . .], we were able to V. . . .
  由於你的 [體貼 / 好意 / 幫忙 / 無私……]，我們得以……

- Words cannot convey my gratitude / gratefulness / appreciation.
  用千言萬語也無法表達我的感激之情。

- You have no idea what this means to me.
  你無法想像這對我的意義有多重大。

- I want to thank you on behalf of (someone) for . . .
  我要替（某人）謝謝你……

- Your [professionalism / leadership / dedication / diligence . . .] has been appreciated / recognized.
  你的 [專業精神 / 領導能力 / 奉獻 / 勤勉……] 為大家所賞識。

A. Hospitality 殷勤招待

TO: Jack Huang
FROM: Sorin Dogaru
DATE: Feb. 18th
SUBJECT: Thanks for Everything
ATTACHMENT: Trip to Taiwan (47.7 KB)

Dear Mr. Jack Huang:

We are very **grateful**[2] for your hospitality during our visit to Taipei. We all enjoyed seeing the beautiful and interesting sights, especially the National Palace Museum and Taipei 101. Thank you very much for sharing the unique culture and history of Taiwan with us. The food was definitely one of the highlights of the trip. We feel very fortunate to have tasted many traditional Taiwanese **delicacies**.[3] Owing to your thoughtfulness, we were also able to visit all kinds of shops — my wife really loves to shop.

Our stay was truly memorable and will not be soon forgotten. We only hope to have the occasion to return your kind **generosity**.[4] By the way, I've attached some of the photographs we took. I hope you like them.

Words cannot convey my gratitude.

Best regards,
Sorin Dogaru

收件人：黃傑克
寄件人：索林・多加魯
日期：二月十八號
主旨：感謝一切
附件：台灣之旅（47.7 KB）

親愛的黃傑克先生：

非常感激您在我們拜訪台北時的熱情款待。我們都很享受參觀那些美麗又有趣的景點，特別是國立故宮博物院和台北 101。很感謝您和我們分享台灣獨特的文化和歷史。台灣的美食絕對是這趟旅行最精采的部分之一。我們覺得非常幸運能夠品嚐到許多傳統的台灣佳餚。由於您的體貼，我們還得以逛逛各式各樣的商店 — 我太太真的很愛購物。

這段停留的日子非常難忘，將久久無法忘懷。我們但願將來能有機會可以回報您的慷慨大方。順帶一提，我附上一些我們拍的照片，希望您會喜歡。

千言萬語也無法表達我的感激之情。

謹上
索林・多加魯

Vocabulary and Phrases

1. **hospitality** [ˌhɑspɪˋtælətɪ] *n.* 殷情招待；好客
   Thanks to your hospitality, we had a nice stay in San Francisco.
   由於你的殷情招待，我們的舊金山之行很愉快。

2. **grateful** [ˋgretfəl] *adj.* 感激的
   They said they were grateful for all the help they received.
   他們說很感激所有得到的幫助。

3. **delicacy** [ˋdɛlɪkəsɪ] *n.* 佳餚；美味
   Fried crickets are considered a delicacy in Thailand.
   炸蟋蟀在泰國被視為美味佳餚。

4. **generosity** [ˌdʒɛnəˋrasətɪ] *n.* 慷慨；寬宏大量
   Uncle Lynn was praised by many for his generosity.
   很多人讚揚林恩叔叔的慷慨。

5. **aroma** [əˋromə] *n.* 香味；香氣
   What I love the most about coffee is the aroma.
   我最愛咖啡的香氣。

6. **cheongsam** [ˋtʃɔŋˌsam] *n.* 旗袍；長袍馬褂
   The flight attendant wore a beautiful cheongsam.
   那位空服員穿了一件漂亮的旗袍。

B. Receiving Gifts 接受禮物

To: Frank Cheng
From: Rick Jones
Date: August 21
Subject: Your Gifts and Your Time

Dear Mr. Cheng,

Thank you for the gifts. You have no idea what this means to us. But you really shouldn't have. You have already done so much.

First, I want to thank you for the Oolong tea. The box is very elegant and the tea is wonderful. What a truly unique **aroma!**[5] My wife wants me to tell you the **cheongsam**[6] is **stunning,**[7] and I couldn't agree more. She's going to save it for special occasions. Finally, I want to thank you **on behalf of**[8] my colleagues for the "I Love Taiwan" T-shirts. They are very cute.

Once again, thanks for not only taking the time to show us around, but to send us gifts as well.

Yours truly,
Rick Jones
Vice President
Cleary and Associates, Ltd.

中譯

收件人：鄭法蘭克
寄件人：瑞克‧瓊斯
日期：八月二十一號
主旨：你的禮物和寶貴時間

親愛的鄭先生：

謝謝您的禮物。您無法想像這對我們的意義有多重大。不過您真的不用那麼客氣。您已經為我們做很多了。

首先，我想謝謝您的烏龍茶。那個盒子非常漂亮、茶也很棒，它的香氣真的很獨特！內人要我轉達她的旗袍非常漂亮，我也這麼認為。她準備要留到特別的場合再穿。最後，我想要替我的同事們感謝您送的那些印有「I Love Taiwan（我愛台灣）」的 T 恤。它們非常可愛。

再一次感謝您花時間帶我們四處參觀，並送我們禮物。

謹啟
瑞克‧瓊斯
副總裁
Cleary and Associates, Ltd.

---

7. stunning [ˈstʌnɪŋ] *adj.*
（口語）非常漂亮的；極好的

The view from the balcony was stunning.
從陽台看出去的景色非常漂亮。

8. on behalf of 代表

He spoke on behalf of his company.
他代表他的公司發言。

Language Corner

1. Returning Hospitality

We only hope to have the occasion to return your kind generosity.

You Might also Say . . .

- We look forward to returning the favor.
- We hope to repay your kindness in full.
- Next time, everything is on us.

2. Agreeing with Someone

I couldn't agree more.

You Might also Say . . .

- I agree with you completely.
- I'm on your side 100 percent.
- I see things the same way.

C. Hard Work 辛勤工作

To: Steve Ryder
From: Richard Barlow
Date: June 10
Subject: Much Appreciated

Dear Steve,

First of all, I would like to take this opportunity to express my **heartfelt**[1] appreciation to you for your hard work and assistance in making the project successful.

The project's outcome is the direct result of your leadership and planning. Your technical skill and knowledge of both product design and the manufacturing process have been appreciated by everyone here for a long time. Initial **feedback**[2] on our newest products has been very encouraging. **Along the same vein**,[3] our order volume is increasing on a daily basis. Furthermore, the production line is **at full capacity**[4] and the line is running **shifts**[5] **around the clock**.[6]

Again, thank you so much for your **enthusiastic**[7] participation in our research. This project would not have **come to fruition**[8] without your professional advice and kind assistance.

Drop by and visit us whenever you are in Taiwan. When I get to the US, I hope we can get together.

Yours sincerely,
Richard Barlow
Managing Director
Maples Taiwan

中譯

收件人：史提夫・萊德
寄件人：理查・巴洛
日期：六月十號
主旨：萬分感激

親愛的史提夫：

　　首先，我想藉此機會向你表達我由衷的感謝，謝謝你努力工作並提供協助讓這個案子得以成功。

　　這個案子的成果要直接歸功於你的領導和規劃。長久以來這裡的每個人都很欣賞你在產品設計和製造過程方面的技術和知識。我們新產品的初步反應非常鼓舞人心。同樣地，我們的訂單量每天都在增加。此外，生產線達到了最高產能，日以繼夜輪班生產。

　　再一次多謝你熱心地參與我們的調查研究。沒有你的專業建議和親切協助，這個案子是無法完成的。

　　你任何時候到台灣來，都歡迎到我們這裡走走看看。我去美國的時候，希望我們可以聚一聚。

謹上
理查・巴洛
總經理
Maples 台灣分公司

Vocabulary and Phrases

1. heartfelt [ˈhɑrtˌfɛlt] *adj.* 衷心的；真誠的

Your coworker's apology seemed very heartfelt.

你同事的道歉似乎很真誠。

2. feedback [ˈfidˌbæk] *n.* 反應；意見

The feedback they've been receiving on their new product has been helpful.

他們收到有關新產品的反應很有幫助。

3. along the same vein 同樣地

We need to get sales up; along the same vein, we must also provide better customer service.

我們得提高銷售量；同樣地，我們也必需提供更好的顧客服務。

4. at full capacity 達到最高產能

The most disappointing thing is that the factory has never operated at full capacity.

最令人失望的是工廠從未達到最高產能。

5. shift [ʃɪft] *n.* 輪班；輪班工作時間

Doris hates to work the night shift.

多麗思討厭值夜班。

6. around the clock 日以繼夜地

Construction has been going on around the clock.

建築工程正日以繼夜地進行中。

7. enthusiastic [ɪnˌθjuziˈæstɪk] *adj.* 熱心的；熱烈的

We're enthusiastic about the changes that are being made.

我們很熱衷於所實施的變革。

8. come to fruition 實現；完成

Unfortunately, she died before her plan could come to fruition.

很遺憾地，她在計畫完成前就過世了。

Language Corner

1. Giving Credit

The project's outcome is the direct result of your leadership and planning.

You Might also Say . . .

- Your direction and vision made the project successful.
- Your guidance and foresight resulted in the project being a success.

2. Getting the Job Done

This project would not have come to fruition without your professional advice and kind assistance.

You Might also Say . . .

- This project would not have been accomplished without your suggestions and cooperation.
- This project would not have seen the light of day without your input and collaboration.

Unit Three
Complaints 抱怨

抱怨信函需先表明抱怨的事項，內容力求簡明扼要，語氣堅定而平和，切勿口出惡言，畢竟彼此之間還有業務往來的關係。信函要指名寫給特定對象，才會有專人負責處理。如果沒有得到回音或適當的處理，可再發一封措辭較強硬的信給對方。若問題仍未獲解決，則可改發抱怨信函給更高階的主管。

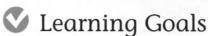 Learning Goals

學習用不同語氣撰寫抱怨信函

Before We Start 實戰商務寫作句型

Topic: Making Complaints 抱怨

◆ I don't think this is appropriate / suitable / fitting because . . .
我認為這不恰當，因為……

◆ I believe you and I discussed this before and agreed that it wouldn't and shouldn't happen / occur / take place.
我相信我們之前討論過此事，並同意它不會也不應該發生。

◆ I wish I would have been notified / informed / advised of this issue / problem / matter (a period of time) ago.
我但願我在（一段時間）前就被告知這個問題。

◆ My experience with . . . so far hasn't been positive / good at all.
到目前為止，我與……接觸的經驗非常不好。

◆ . . . has totally / completely / entirely neglected / disregarded my [attempts / pleas / efforts . . .] to V. . . .
……完全忽視我……的 [嘗試 / 請求 / 努力……]。

◆ I would appreciate it / be thankful / be grateful / be obliged if you could . . .
如果你能……，我會很感激的。

◆ Please be forewarned / alerted / cautioned that you are not to V. without our [representative present / prior approval / written permission . . .].
請注意如果沒有我們的 [代表在場 / 事先同意 / 書面許可……]，你不能……

◆ You can expect / count on [serious consequences / disciplinary action / a severe penalty . . .] if S. + V. . . .
如果……，你會有 [嚴重的後果 / 懲戒處分 / 嚴峻的處罰……]。

A. Communication Failure 溝通不良

To: Ruby
From: Richard
Date: 2/24
Subject: Pickup Confusion

Ruby,

I called Taiwan on my day of departure (2/20) and discovered that the fax I asked you to send on 2/18 with my itinerary and request for pickup arrangements never arrived in Taiwan.

Things worked out OK with some last-minute **scurrying**,[1] but I thought you'd probably want to know that the system didn't work that day. I never did find out exactly what went wrong; perhaps you can look into it and get back to me. I apologize in advance if the **miscommunication**[2] was due to any error on my part.

Later,
eom

中譯

收件人：茹比
寄件人：理查
日期：二月二十四號
主旨：接送烏龍

茹比：

　　我在出發日（二月二十號）打電話回台灣，發現我請妳在二月十八號傳我的行程表和要求安排接送的傳真並沒有傳到台灣。

　　最後在一陣手忙腳亂後，事情順利解決了。但我想或許妳會想知道當天的溝通機制運作不良。我還不知道究竟是哪裡出了錯；或許妳可以查一下再回覆我。如果是因為我這邊的失誤造成傳達不清，我先向妳道歉。

再聯絡
完畢

Vocabulary and Phrases

1. scurrying [ˋskɝɪŋ] *n.* 手忙腳亂；急趕

After some last-minute scurrying, I was able to get his birthday present.
在一陣手忙腳亂之後，我終於買到他的生日禮物。

2. miscommunication [ˌmɪskəˌmjunəˋkeʃən] *n.*
傳達不清；傳達錯誤

I'm sorry about our miscommunication the other day.
我很抱歉我們那天傳達不清。

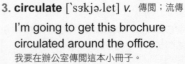

3. circulate [ˋsɝkjəˌlet] *v.* 傳閱；流傳

I'm going to get this brochure circulated around the office.
我要在辦公室傳閱這本小冊子。

4. notify sb of sth 告知某人某事

I never notified David of the errors in his letter.
我從未告知大衛他信裡的錯誤。

B. Mishandling of Faxes 未妥善處理傳真

To: Tony
From: Richard
Date: 12/3
Subject: Fax Problems

Dear Tony,

　　Wendy tells me that usually the AECO and FETS faxes get copied and **circulated**[3] throughout the company with the MCUS faxes. I don't think this is appropriate because many of the faxes I send contain sensitive information, and I believe you and I discussed this before and agreed that it wouldn't and shouldn't happen. I wish I would have been **notified of**[4] this issue a few weeks ago. Before I send any more AECO / FETS faxes, I'll wait for word from you that a system is in place to **route**[5] these faxes to you alone. Thanks.

Best regards,
eom

中譯

收件人：東尼
寄件人：理查
日期：十二月三號
主旨：傳真問題

親愛的東尼：

　　溫蒂告訴我通常 AECO 和 FETS 的傳真會和 Maples 美國總公司的傳真一起備份並在公司傳閱。我認為這並不恰當，因為很多我傳的傳真都包含敏感的資訊。而我相信我們之前討論過此事，並同意它不會也不應該發生。我希望我早在幾週前就已被告知這個問題。在我傳送更多 AECO 或 FETS 的傳真之前，我會先等你通知我，確定給你的這些傳真只有你收得到的機制已經建立了。謝謝。

謹上
完畢

5. **route** [rut] *v.* 按特定路線運送、寄發

All quality-control complaints will be routed to your department.
所有有關品管的抱怨都會傳送到你的部門。

Language Corner

Awaiting a Solution

✦ I'll wait for word from you that . . .
✦ I'm anxiously awaiting your response telling me that . . .
✦ I'll be expecting your swift reply informing me that . . .

For example:

1. I'll wait for word from you that you've taken care of this matter.
2. I'm anxiously awaiting your response telling me that you've bought the necessary materials.

C. Unprofessional Behavior 不專業的行徑

To: Roy
From: Richard
Date: 11/9
Subject: MicroSensation Employee Conduct

Dear Roy,

My experience with MicroSensation so far hasn't been **positive**[1] at all. David Lake sent me a very incomplete package of the touch-screen controllers to quote, and has totally neglected my **subsequent**[2] attempts to get more information. I get the impression that he doesn't really value our partnership.

I believe that personal contact and mutual respect are very important while establishing a new business relationship. David hasn't **demonstrated**[3] much of either. I would appreciate it if you could meet up with him while you're in the States next week and explain how we do business. Hopefully, you and he can also discuss his controller needs to help speed up the process of finishing the quote.

Best regards,
eom

收件人：羅伊
寄件人：理查
日期：十一月九號
主旨：MicroSensation 員工的行為

親愛的羅伊：

到目前為止我與 MicroSensation 接觸的經驗非常不好。大衛‧雷克寄給我一個有關觸控式螢幕控制器但資料非常不完整的包裹要我報價，我後續試圖請他提供更多資訊，他也全然不予理會。我認為他並不是很重視我們的合作關係。

我相信在建立一個新的商業關係時，私人接洽和相互尊重是非常重要的，但大衛在這兩方面的表現都相當欠缺。下週你去美國時如果可以和他碰面，並解釋我們是如何做生意的，我會很感激。但願你們也能討論一下他對控制器的需求，以便加速完成報價的程序。

謹上
完畢

Vocabulary and Phrases

1. **positive** [`pɑzətɪv] *adj.* 正面的；積極的

   The viewers' response to the movie was not positive.
   觀眾對於那部電影的反應並不好。

2. **subsequent** [`sʌbsɪ.kwɛnt] *adj.* 後續的；隨後的

   All subsequent calls will be forwarded to my secretary.
   之後所有的電話都會轉接給我的秘書。

3. **demonstrate** [`dɛmən.stret] *v.* 展現；證明

   He needs to demonstrate that he's ready for a promotion before I give him one.
   在我晉升他之前，他需展現出自己已經做好升職準備了。

4. **apparently** [ə`pɛrəntlɪ] *adv.* 顯然地

   Apparently, we need to hire more staff.
   顯然我們需要聘請更多員工。

5. **proprietary** [prə`praɪə.tɛrɪ] *adj.* 私有的；所有人的

   Any design you make as our employee is proprietary.
   身為我們員工所構思的任何設計，其所有權都屬於公司。

D. Policy Issues 政策問題

To: David Chou
From: Richard Barlow
Date: 10/4
Subject: Company #3C Factory Visit

Mr. Chou,

I was informed that yesterday you took Pai Lee, Quality Manger of Company #3C, on a tour of the GasPack® production line with no Maples representative present. **Apparently**,[4] you need to be reminded of our company's policy.

We consider the GasPack® production line to be **proprietary**.[5] Please be **forewarned**[6] that you are not to show our production lines to anyone without a Maples employee present. You can expect serious **consequences**[7] if you violate policy again. I trust that in the future you will show Maples Corporation the same **courtesy**[8] you give to your other customers.

Regards,
Richard

中譯

收件人：周大衛
收件人：理查・巴洛
日期：十月四號
主旨：#3C 公司到工廠參觀

周先生：

我被告知昨天你在沒有 Maples 公司代表在場的情況下，帶 #3C 公司的品質經理李派參觀 GasPack® 產品的生產線。顯然你需要被提醒有關我們公司的政策。

我們認為 GasPack® 生產線是屬於私有財產的。請注意如果沒有 Maples 公司的代表在場，你不能帶任何人參觀我們的生產線。如果你再次違反規則，會有嚴重的後果。我相信往後你對 Maples 公司也會像你對其他客戶一樣尊重。

祝 商安
理查

---

6. **forewarn** [fɔr`wɔrn] *v.* 事先告知；預先警告

He was forewarned that if he arrived late again, he would be fired.
他被事先警告如果他再遲到就會被解雇。

7. **consequence** [`kɑnsə͵kwɛns] *n.* 後果

Every action has consequences.
凡事有因必有果。

8. **courtesy** [`kɜtəsɪ] *n.* 禮貌；好意

I don't think you've been showing us enough courtesy.
我認為你對我們不夠有禮貌。

Language Corner

Complaining about Business Manners

I get the impression that he doesn't really value our partnership.

You Might also Say . . .

- I sense that he doesn't think our business is important.
- I get the feeling that he doesn't really care about doing business with us.
- I get the notion that our company's business is not a priority to him.

Unit Four
Apologizing 道歉

因本身過失而造成對方
權益受損時，需儘快寫
信向對方表示歉意，以
修補雙方的關係。要真
心誠意地道歉，切忌找
理由推卸責任，同時要
具體說明所採取的補救
措施，並承諾類似事件
不會再發生。

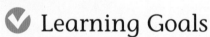 **Learning Goals**

學習撰寫道歉信函

Before We Start 實戰商務寫作句型

Topic: Making an Apology 道歉

* I sincerely / deeply / truly apologize for . . .
  對於……，我誠摯地道歉。

* I am very sorry for / about . . .
  我很抱歉……

* I hope you will excuse / forgive me for . . .
  我希望你能原諒我……

* Another apology is in order for the [communication breakdown / mix-up / blunder . . .] on our end / part / side.
  另外我們這邊 [溝通不良 / 搞混 / 出錯……] 的部分，也要跟你道歉。

* Please accept my apology for any [inconvenience / problem / trouble / annoyance . . .] this [incident / situation / error / oversight . . .] has caused.
  對於這個 [事件 / 情況 / 錯誤 / 疏忽……] 所造成的 [不便 / 問題 / 麻煩 / 不愉快……]，請接受我的致歉。

* Thank you for your [patience / understanding . . .].
  謝謝你的 [耐心 / 諒解……]。

* We're willing to offer you [a discount / a gift certificate / some coupons . . .] as a token of our apology.
  為表示歉意，我們願意提供你 [折扣 / 一張禮券 / 一些折價券……]。

A. Late Response 延誤回信

TO: Kevin Green
FROM: Richard Barlow
DATE: December 22
SUBJECT: Follow-up on Machined Parts*

Dear Kevin,

Thanks for your fax and your interest in Far East Tech. I apologize for not being in contact with you sooner regarding your message. I have been away on business for a **protracted**[1] length of time and got **waylaid**[2] by a typhoon in Hong Kong as I was trying to get home.

As you know, the area of machined parts is one in which we have **vast**[3] experience, and we are, of course, very interested in the business you mentioned. Please do send details and I will be pleased to supply you with any additional information you may need.

Again, I'm very sorry for the delayed response.

Cheers,
Richard Barlow
Managing Director
Far East Technical Services

中譯

收件人：凱文・葛林
寄件人：理查・巴洛
日期：十二月二十二號
主旨：機械零件後續追蹤

親愛的凱文：

謝謝你的傳真和對 Far East Tech 感興趣。我很抱歉無法針對你的來信儘快和你聯絡。我有一段相當長的時間出差在外，在回程時又因颱風而滯留在香港。

如你所知，我們在加工零件這方面非常有經驗。我們當然也對你提到的生意非常有興趣。請務必提供更多詳盡的資料，我也會很樂意提供你可能需要的任何其他資訊。

再次抱歉耽誤回信。

工作愉快
理查・巴洛
總經理
Far East Technical Services

Vocabulary and Phrases

1. protracted [proˋtræktɪd] *adj.* 長時間的；拖延的

I waited at the bus station for a protracted period of time.
我在公車站等了相當長的一段時間。

2. waylay [ˋweˏle] *v.* 耽擱；攔住；伏擊

He was waylaid by a tornado that delayed his flight.
龍捲風使他的班機延誤，他也因此被耽擱了。

3. vast [væst] *adj.* 廣泛的；大量的

The supermarket has a vast selection of products
那間超市有很多不同產品可供選擇。

4. intimate [ˋɪntəˏmet] *v.* 暗示；提示

He intimated that he might sell the company.
他暗示他可能會賣掉這間公司。

5. in order 必要的；適當的

You got the job?! A celebration is in order.
你得到那份工作了？！需要慶祝一下。

 Biz Focus　　　　★ machined part 經由車床加工的機械零件

B. Shipment Confusion 出貨混亂

To: Bob Sanders
From: Richard Barlow
Date: 3/31
Subject: The Fax #AB-123

Dear Bob,

First, let me apologize for my last e-mail. It was obviously unclear. I didn't mean to **intimate**[4] that we've produced 100,000 German version units, but I can see how the message could be read that way. To confirm the current production status, Mica has produced and shipped 100,008 UK version units and 10,000 German version units.

Secondly, another apology is **in order**[5] for the miscommunication on our end. Clearly, the German units should not have been shipped, and no others will be shipped until we receive word from you.

Best regards,
eom

中譯

收件人：鮑伯・桑德斯
寄件人：理查・巴洛
日期：三月三十一號
主旨：編號 AB-123 的傳真

親愛的鮑伯：

首先，讓我針對上一封電子郵件向你道歉，那封信顯然很不清楚。我無意暗示我們已經製造了十萬件德規產品，不過我能理解那封訊息是有可能會被那樣解讀。確認現在的生產現況是：Mica 工廠已經製造並出貨十萬零八件英規產品和一萬件德規產品。

其次，針對我們這邊溝通不良的部分，我也要向你道歉。顯然我們不應該出德規產品，而在接到你們的通知之前，我們將不會再出任何德規產品。

謹上
完畢

Language Corner

1. Introducing Likely Known Info

+ As you know, . . .
+ As you may have noticed, . . .
+ As you might have guessed, . . .
+ You might already be aware that . . .

For example:

1. As you may have noticed, we have a new staff member.
2. You might already be aware that we'll be moving our office next month.

2. Awaiting Permission

No others will be shipped until we receive word from you.

You Might also Say . . .

- We won't ship anything else until we get your go-ahead.
- We'll hold all units until we receive notice of your permission to ship.

C. Misplaced Shipment 貨品遺失

快遞公司將 Maples 美國總公司要寄給台灣分公司的包裹寄丟了,於是寫信向寄件者 Ms. Hoffman 道歉。

February 3

Ms. Shelly Hoffman
Maples Corporation
435 8th Street
Los Angeles, CA

Dear Ms. Hoffman:

The problem **encountered**[1] with your January 27 shipment to Maples Corporation in Taichung, Taiwan on our International Air Waybill* 412-73341341 has been **brought to my attention**.[2]

Please accept my apology for any inconvenience this incident has caused. Our records indicate the on-time arrival of your shipment in our Taipei clearance facility.* Unfortunately, we're unable to inform you of a current package status.

Our primary goal is to handle each package in a professional and **expedient**[3] manner. Your experience with this shipment is neither typical nor is it an accurate reflection of the type of service we wish to provide. Please be assured that management has been advised of your problem.

Because of our **commitment**[4] to customer satisfaction, we do not expect payment for unsatisfactory service. Therefore, your account will be **credited**[5] for the shipment's freight charges. If you should **inadvertently**[6] receive a statement from us, please forward it with a copy of this letter to the address provided on the invoice. Furthermore, all **pertinent**[7] information has been passed on to our claims department for compensation.

Thank you for your patience. We value the trust our customers place in us to handle their worldwide **express**[8] shipments. We're also willing to offer you a 30-percent discount off your next shipment **as a token of**[9] our apology. We look forward to better serving your future needs.

Sincerely,

Eddy M. Harris
Eddy M. Harris
Customer Relations Manager
International Correspondents

 中譯

二月三號

雪莉·霍夫曼女士
Maples 公司
加州洛杉磯市
第八街 435 號

親愛的霍夫曼女士：

我已經注意到您在一月二十七號托運到 Maples 台灣台中分公司，航空運貨單編號 412-73341341 那批貨所遭遇到的問題。

對於此事件造成的不便，請接受我的道歉。我們的記錄顯示您的貨品準時抵達我們在台北的貨物進口清關處。但很遺憾地我們無法向您報告包裹現在的狀況。

我們的首要目標是以專業和便捷的方式來處理每一件包裹。您這次的出貨經驗並不常見，也無法準確反映出我們所期望提供的服務類型。請您放心管理單位都已被告知您的問題。

因為我們承諾要讓顧客滿意，所以不符合要求的服務

我們是不會收費的。因此，這批貨的運費將被退還到您的帳戶。如果您不慎收到我們的帳單，請將它連同本封信的影本一同轉寄到發票上提供的地址。此外，所有相關的資訊也都交給我們的理賠部門來處理補償金事宜。

謝謝您的耐心。我們重視客戶託付我們來處理他們全球快遞貨運運時所給予的信賴。為表達我們的歉意，下一批運貨我們也願意提供您七折的優惠。我們期待能為您未來的需求提供更優質的服務。

祝 商安
艾迪・M.・哈利斯
客服經理
International Correspondents

Vocabulary and Phrases

1. **encounter** [ɪnˈkaʊntɚ] *v.* 遇到（困難、危險等）
   She encountered difficulties while trying to use her credit card online.
   她在網路上刷卡時遇到了困難。

2. **bring sth to sb's attention** 使某人注意到某事
   Thanks for bringing the error to my attention.
   謝謝你讓我注意到這個錯誤。

3. **expedient** [ɪkˈspidɪənt] *adj.* 方便的；權宜之計的
   They identified the problem and tried to find the most expedient solution.
   他們發現了這個問題，並設法找出最好的權宜之計。

4. **commitment** [kəˈmɪtmənt] *n.*
   承諾；保證；承擔的義務
   Every worker needs to make a commitment to excellence.
   每位員工都必需承諾有傑出的表現。

5. **credit** [ˈkrɛdɪt] *v.* 把……記入貸方
   We'll credit your account with the 100-dollar refund.
   我們會將一百元的退款轉入你的帳戶。

6. **inadvertently** [ˌɪnədˈvɝtn̩tlɪ] *adv.* 不慎地；非故意地
   I inadvertently knocked the child over as I walked past.
   我走過去時不小心把那個小孩撞倒了。

7. **pertinent** [ˈpɝtn̩ənt] *adj.* 有關的；相干的
   She didn't mention that issue because she didn't think it was pertinent.
   她沒有提那件事因為她認為它並不相關。

8. **express** [ɪkˈsprɛs] *adj.* 快遞的；快速的
   We decided to take the express train to the next city.
   我們決定搭快車到下一個城市。

9. **as a token of** 作為……的象徵、標誌
   He gave his girlfriend's parents a gift as a token of his respect.
   他送給他女友的父母一個禮物以表示尊重。

Biz Focus

★ **air waybill** 航空運貨單（縮寫為 AWB）

★ **clearance facility** 貨物進口清關處

Language Corner

Easing a Customer's Worries

Please be assured that management has been advised of your problem.

> *You Might also Say . . .*

- You can be confident that the proper people have been notified of the situation.
- Please be secure in knowing that this issue has been forwarded to the appropriate party.

Unit Five
Welcome Aboard 迎新

藉由 welcome letter 來歡迎公司的新同事，可以讓新進人員感覺受到重視和賞識，進而增加對公司的歸屬感，也可藉此機會讓對方認識自己，幫助雙方儘早建立起工作情誼。

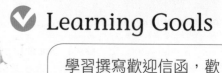
✔ Learning Goals

學習撰寫歡迎信函，歡迎新同事加入

Before We Start 實戰商務寫作句型

Topic: Welcoming a Newcomer 歡迎新成員

◆ Please help me in welcoming (someone) to (company).
請幫我一起歡迎（某人）加入（某公司）。

◆ I hope you'll all join me in giving a warm welcome to (someone).
希望大家和我一起熱烈歡迎（某人）。

◆ Let's all welcome (someone) aboard.
我們歡迎（某人）加入。

◆ I am very pleased / excited / glad you have [joined our team / decided to work with us / chosen to become a part of the company . . .].
我很高興你 [加入我們的團隊 / 決定與我們共事 / 選擇成為公司的一分子……]。

◆ I'll give you a quick rundown / summary / outline of the [orientation process / company code / training schedule . . .].
我將向你很快簡述 [新進員工訓練的過程 / 公司規範 / 訓練時間表……]。

◆ For starters / To begin with / To start off, you'll need to fill out a series of / a set of / an assortment of forms.
首先，你將必需填寫一系列 / 各種表格。

◆ I'll [brief you on / teach you about / introduce you to . . .] our [procedure(s) / strategy / protocol . . .] for . . .
我將 [和你簡報 / 教你 / 向你介紹……] 我們……的 [程序 / 策略 / 禮儀……]。

◆ If there's anything we can do to help you settle into / get used to / get accustomed to your new position / post / job, just let us know.
如果有什麼是我們可以做以幫助你適應新職務的，就讓我們知道。

A. Introducing a New Employee 介紹新成員

To: all@mat.com
From: Richard
Date: 3/24
Subject: New Colleague

Dear All,

Please help me in welcoming Lisa Yang to Maples Corporation. Lisa joined the MAT family yesterday as Secretary to the Vice President.

Lisa graduated **with honors**[1] from a local university with a BA degree in English, and has six years of international trading experience. Her product experience has been with toys, bicycles, and hand tools. She has also attended **quite a few**[2] trade shows in the States and Europe, working booths as her company's representative. I'm confident that she will be a valuable contributor to our company.

Richard

中譯

收件人：all@mat.com
寄件人：理查
日期：三月二十四號
主旨：新同事

親愛的各位：

請幫我一起歡迎楊麗莎加入 Maples 公司。麗莎昨天加入 MAT 這個大家庭，擔任副總裁的秘書。

麗莎以優異的成績畢業於一所本地大學，並取得英文學士學位；同時她也有六年國際貿易的經驗。她所接觸的產品經驗包含玩具、腳踏車和手動工具。她也曾代表她的公司參加相當多在美國和歐洲舉辦的商展。我有信心她會為我們公司帶來有價值的貢獻。

理查

Vocabulary and Phrases

1. with honors 以優異成績

He earned his degree with honors from American College in 1997.
他在 1997 年以優異的成績畢業於美國大學。

2. quite a few 相當多

I'm not from New York, but I know quite a few people in the city.
我不是來自紐約，但我認識相當多那裡的人。

3. orientation [ˌɔriənˈteʃən] *n.*
（針對新職員的）情況介紹（orient *v.*）

I'll e-mail you a copy of the orientation schedule.
我會用電子郵件寄一份新進員工訓練的時間表給你。

4. get adjusted to 適應；熟悉

It takes small children a while to get adjusted to new people.
小孩子需要花一段時間來熟悉陌生人。

5. rundown [ˈrʌnˌdaun] *n.* 概要

She was late for the meeting, so her coworker gave her a rundown of what had been said.
她開會遲到了，所以她的同事向她概述剛才講過的事情。

6. fill out 填寫

He decided to fill out an application for the job.
他決定要填寫申請表應徵那份工作。

B. Orienting New Staff 使新進員工適應環境

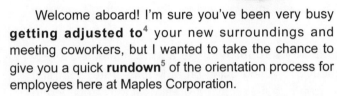

To: Lisa Yang
From: Richard Barlow
Date: 3/25
Subject: Job **Orientation**[3]

Dear Lisa,

Welcome aboard! I'm sure you've been very busy **getting adjusted to**[4] your new surroundings and meeting coworkers, but I wanted to take the chance to give you a quick **rundown**[5] of the orientation process for employees here at Maples Corporation.

For starters, you'll need to **fill out**[6] a series of forms. That way, we can get you set up with a parking pass, an ID card, health insurance, and the direct deposit process. You can pick up the forms from my assistant Vincent.

I'll also **swing by**[7] your desk this week and brief you on our procedures for answering phones and taking messages.

Cheers,
Richard

中譯

收件人：楊麗莎
寄件人：理查・巴洛
日期：三月二十五號
主旨：認識新工作環境

親愛的麗莎：

歡迎加入！我相信妳現在正忙著適應新環境和認識同事，但我想要趁這個機會很快地向妳簡述 Maples 公司新進員工的訓練過程。

首先，妳將必需填寫一系列的表格。那樣我們才能幫妳辦理停車證、識別證、健保和薪資轉帳。妳可以跟我的助理文森拿那些表格。

這星期我也會過去妳的位子，向妳簡報我們接電話和記錄留言的程序。

工作愉快
理查

---

7. swing by（中途）短暫停留

She had to swing by her friend's house to drop off the MP3 player.
她中途必需拿 MP3 播放器到她朋友家。

Language Corner

1. Expressing Confidence in a New Employee

I'm confident that she will be a valuable contributor to our company.

You Might also Say . . .

- I'm certain that she will be a positive addition to our company.
- She will undoubtedly be a key member of our company.

2. Making Arrangements for Newcomers

| get sb set up / started (with sth) | 替某人辦理、取得…… |
| get sth up and running | 使某物開始運作 |

For example:

1. He's going to get the new worker set up with a swipe card for the elevator.

2. We need to get her computer system up and running.

C. Extending a Welcome 歡迎新人

To: Jack Huang
From: Tony Hughes
Date: August 20
Subject: Welcome to the Staff

Dear Jack,

I want to take this opportunity to welcome you and let you know that I am very pleased you have joined our team.

I believe you will find Richard very **supportive**.[1] He always makes sure new employees are **up to speed**[2] about everything going on at Maples. After your first day, you will probably already be aware that we have a great deal to correct and to accomplish, and it will not necessarily be easy. Our opportunities will only improve, especially if we can get better **organized**[3] and **stick to**[4] our business plan. The **vision**[5] that our CEO, Ben, has developed for our company is very sound. As we discuss it with our current and **potential**[6] customers, they often confirm that we are headed in the right direction.

There is one thing you can always count on while working for this company: "Opportunity to grow as an **individual**[7] is yours for the taking." Richard will give you as much opportunity and responsibility as you are willing to take.

Again, I wanted to give you a warm welcome. If there's anything we can do to help you settle into your new position, just let us know. I am looking forward to working with you, and I feel confident that you will be able to step in and make an **impact**[8] right away.

Sincerely,
Tony Hughes
President of Maples Corporation

中譯　收件人：黃傑克
寄件人：東尼·休斯
日期：八月二十號
主旨：歡迎成為我們的一分子

親愛的傑克：

　　我要藉此機會歡迎你，並讓你知道我很高興你能加入我們的團隊。

　　我相信你將發現理查會非常支持你。他總是能確保讓新進員工了解 Maples 公司的最新現況。在你上完第一天班後，你或許就將注意到我們有很多部分尚待修正和完成，而這不盡然是一件簡單的事。我們將有很好的機會，特別是如果我們能更有組織且恪守我們的營運計畫。我們的執行長班所開創的願景十分完善，當我們與目前和潛在的客戶討論時，他們也時常肯定我們走的方向是正確的。

在這裡工作，有一件事是你可以一直期待的：「每個人都有發展的機會」。只要你願意，理查會儘可能給你機會和責任。

讓我再一次向你表示熱烈歡迎。如果有什麼是我們可以做以幫助你適應新職務的，就讓我們知道。我期待與你共事，我有信心你的加入必定能很快對公司有所貢獻。

祝 商安
東尼·休斯
Maples 公司總裁

Vocabulary and Phrases

1. supportive [sə`pɔrtɪv] *adj.* 支持的

His wife was very supportive when he was upset.
他老婆在他沮喪的時候非常支持他。

2. up to speed 了解最新狀況

Karen is in charge of bringing the newcomer up to speed about our products.
凱倫負責幫助那位新進員工了解我們公司產品的最新狀況。

3. organized [`ɔrgə͵naɪzd] *adj.* 有組織的；有系統的

She needs to get organized if she wants to succeed in this job.
如果她這份工作要成功，就必需有條不紊。

4. stick to 忠於；信守

In business, you should always stick to what you do best.
做生意應該要一直忠於自己最拿手的。

5. vision [`vɪʒən] *n.* (預想的)前景；展望

The presidential candidate shared his vision for the country's future.
那位總統候選人分享了他對國家未來的願景。

6. potential [pə`tɛnʃəl] *adj.* 潛在的；可能的

The three potential employees are hoping they'll be hired.
那三位候選的員工都希望他們能被雇用。

7. individual [͵ɪndə`vɪdʒuəl] *n.* 個人；個體；人物

He doesn't like group projects because he works better as an individual.
他不喜歡團隊計畫，因為他比較擅長一個人工作。

8. impact [`ɪm͵pækt] *n.* 影響

The new product made a positive impact on the company's profits last month.
這個新產品對公司上個月的獲利有正面的影響。

Language Corner

1. Making Something Available

+ be yours for the taking
+ be up for grabs
+ be open to everyone
  人人有份

For example:

1. The promotion is yours for the taking if you keep working hard.
2. The food in the lunchroom is up for grabs.

2. Awaiting a Future Event

I am looking forward to + V-ing
I can't wait to V.
I'm anxiously anticipating + V-ing

我迫不及待……

For example:

1. I can't wait to start working here.
2. I'm anxiously anticipating meeting my new colleagues.

Unit Six
Farewells 道別

自己或別人要離職時，都可以寫道別信函。自己要離職時，信的開頭要說明離開的日期，不必詳述離職原因。不管是自己或對方要離職，都要感謝對方在工作上所做的協助，祝福對方未來一切順心如意，並留下日後聯繫的資料。

✔ Learning Goals

1. 學習於離職時撰寫道別信函

2. 寫信向即將離職的同事致意

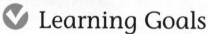

Before We Start 實戰商務寫作句型

Topic 1: When You Quit . . . 自己辭職

◆ I will be leaving / resigning from / stepping down from my position as . . .
我將辭去⋯⋯的職務。

◆ I am grateful for [your kind assistance / the tremendous amount of support you've given me / the opportunity to work with you . . .] over the years.
我很感謝這些年來 [你親切的協助 / 你給予我大量支持 / 與你共事的機會⋯⋯]。

◆ I've enjoyed / taken great pleasure in working with you and respect [the great job you do / your dedication / your hardworking nature . . .].
我很高興能與你共事，並敬重 [你的工作成就 / 你的奉獻 / 你努力工作的本質⋯⋯]。

Topic 2: When a Coworker Is Leaving . . . 同事離職

◆ I'm sorry to hear that you're leaving / quitting / resigning.
我很遺憾聽到你要離職。

◆ I wish you [the best of luck / success / much prosperity . . .] in your [new entrepreneurial endeavor / future pursuits / next job . . .].
我祝你 [創業 / 未來的事業 / 下一份工作⋯⋯] [順利 / 成功 / 鴻圖大展⋯⋯]。

◆ You've always been [my mentor / a valuable member of our team . . .].
你一向都是 [我的良師益友 / 我們團隊中很有價值的成員⋯⋯]。

◆ Please let me know what your new contact information will be so that we can keep in touch / stay connected.
請告訴我你新的聯絡方式，這樣我們才能保持聯繫。

◆ (Company) is deeply appreciative of / obliged to all your [hard work / professionalism / loyalty . . .] for the past (number) years.
（某公司）十分感激你過去（幾）年來的 [努力工作 / 專業精神 / 忠誠⋯⋯]。

◆ Your [consideration / effort / perseverance . . .] has always [impressed us / served as an example for us . . .].
你的 [體貼 / 努力 / 毅力⋯⋯] 向來 [讓我們留下深刻印象 / 是我們的模範⋯⋯]。

A. Signing Off 離職

To: All MAT Employees
From: Andy Anderson
Date: 8/24
Subject: Moving On

Dear Friends & Colleagues,

As you may already know, effective September 1st, I will be stepping down from my position as Office Manager for Maples Taiwan. Next month, I will **relocate**[1] to San Diego and **embark upon**[2] my new job as Sales Manager for DTG Tech, a manufacturer which produces high-tech **appliances**.[3] I am grateful for having had the opportunity to work closely with all of you at Maples Corporation and for your kind **assistance**[4] through the years. I wish you the best for a very bright future!

If you need to reach me, my new contact information is:
238 Clark Street
San Diego, CA 91147
619-376-0954

Best regards,
Andy Anderson
Andy Anderson
Office Manager
Maples Taiwan

中譯

收件人：MAT 全體員工
寄件人：安迪‧安德森
日期：八月二十四號
主旨：離職

親愛的朋友和同事們：

你們可能已經知道了，我將辭去 Maples 台灣分公司辦公室經理的職務，從九月一號生效。下個月我將會搬到聖地牙哥，並開始我在 DTG Tech（一間高科技設備製造商）出任業務經理的新工作。我很感激能有機會與你們在 Maples 共事，也謝謝你們這些年來親切的協助。祝你們的前途一片光明！

如果你們需要聯絡我，我新的聯絡方式是：
91147 加州聖地牙哥市
克拉克街 238 號
619-376-0954

謹上
安迪‧安德森
辦公室經理
Maples 台灣分公司

Vocabulary and Phrases

1. relocate [ri`loˌket] *v.* 重新安置

The family plans to relocate to Santa Fe in the spring.
那一家計畫在春天的時候搬到聖塔菲。

2. embark upon 著手；從事

She will embark upon a new career as a dentist.
她將要開始她當牙醫的新事業。

3. appliance [ə`plaɪəns] *n.* 設備；用具

The washer is the most expensive appliance in my house.
那台洗衣機是我家最貴的設備。

4. assistance [ə`sɪstəns] *n.* 協助

At first, the new employee needed a lot of assistance.
一開始那位新職員需要很多的協助。

5. take pleasure in 以……為樂

You shouldn't take pleasure in the bad luck of others.
你不應該以別人的不幸為樂。

6. bid [bɪd] *v.* 表示（問候）

I will bid her farewell at her good-bye party.
我將在她的歡送會上向她道別。

B. Thanking Coworkers 感謝同事

TO: all@mat.com
FROM: Jack Huang
DATE: 5/15
SUBJECT: Farewell Party

Dear Coworkers,

Thank you, my friends, for your warm and generous farewell at the restaurant last night. I had a great time and I enjoyed the chance to spend some extra time with all of you before I move on. I have **taken** great **pleasure in**[5] working with you and respect the great job you do for MAT. It's a small world; our paths may very well cross again. Anyway, I will be here until the end of next week. I will **bid**[6] you a final farewell then.

Thanks again for **throwing**[7] me a party. It was very **considerate**[8] of all of you.

Best wishes,
Jack

中譯

收件人：all@mat.com
寄件人：黃傑克
日期：五月十五號
主旨：歡送派對

親愛的同事們：

　　朋友們，謝謝大家昨晚在餐廳舉辦溫馨又豐盛的送別會。我玩得很開心，也很高興能有這個機會，可以在我離開前和你們大家另外聚一聚。我很高興能與你們共事，也很敬佩你們在 MAT 的傑出表現。這世界很小；也許我們會再見面。無論如何，我會一直待到下個星期結束。到時候我會向你們作最後道別。

　　再一次謝謝你們為我舉辦派對。你們大家真的很體貼。

謹上
傑克

---

7. **throw** [θro] v.（口語）舉行；舉辦

This Friday, we're going to throw a surprise party.
這個星期五我們要舉辦一個驚喜派對。

8. **considerate** [kən`sɪdərɪt] adj. 體貼的；考慮周到的

The coworker was very considerate and always went outside to smoke a cigarette.
那位同仁很體貼，都會到室外抽煙。

Language Corner

Saying You Might Meet Again

It's a small world; our paths may very well cross again.

You Might also Say . . .

- We might run into each other sometime in the future.
- Who knows? Maybe our paths will intersect in the coming days.

C. Wishing Someone Well 獻上祝福

To: Monica Bennett
From: Todd Andrews
Date: November 20
Subject: Re: Farewell

Monica,

I'm sorry to hear that you're leaving, Monica, but I wish you the best of luck in your new **entrepreneurial**[1] **endeavor!**[2] Thank you for the **tremendous**[3] amount of support you've given me over the years. You've always been my mentor. I wouldn't be where I am today without all of your help.

I am glad that you will be able to grab a piece of the growing Asian travel market. I believe you have picked a good **field.**[4]

I'd love to visit you on one of my trips to Taichung. Please let me know what your new e-mail address and phone number will be so that we can keep in touch.

Ciao,
Todd

中譯

收件人：莫妮卡‧班奈特
寄件人：陶德‧安德魯斯
日期：十一月二十號
主旨：回覆：道別

莫妮卡：

　　莫妮卡，聽到妳要離職我感到很遺憾，但我衷心祝福妳順利開創新事業！感謝妳這些年來給予我大量支持。妳一向都是我的良師益友，沒有妳的幫助我不會有今天的成就。

　　我很高興妳可以分到一塊正在逐漸成長的亞洲旅遊市場這塊大餅。我相信妳選對了行業。

　　到台中時我很想去拜訪妳。請告訴我妳新的電子郵件地址和電話號碼，這樣我們才能保持聯繫。

再聯絡
陶德

Vocabulary and Phrases

1. entrepreneurial [ˌɑntrəprəˋnjʊrɪəl] *adj.*
創業者的；企業家的

He's always searching for new entrepreneurial opportunities.
他一直在尋求新的創業機會。

2. endeavor [ɪnˋdɛvɚ] *n.* 努力

Tony will be very successful in his new business endeavor.
東尼的新事業將會非常成功。

3. tremendous [trɪˋmɛndəs] *adj.* 極大的

Your contribution to this project has been a tremendous help to me.
你對這個案子的貢獻真的給了我很大的幫助。

4. field [fild] *n.* 領域；範疇

Edison is specialized in the field of biotechnology.
艾迪森的專長在生物科技領域。

5. dedication [ˌdɛdəˋkeʃən] *n.* 奉獻

He was recognized for his decades of dedication to teaching.
他對教學數十年的奉獻獲得了表揚。

6. fill sb's shoes 接替某人的職務

They are trying to find someone to fill the CEO's shoes.
他們正試著找人來接替執行長的職務。

D. Farewell Memo 道別短箋

Good-bye and Good Luck
Jack Huang

Maples Corporation is deeply appreciative of all your hard work and **dedication**[5] for the past twelve years. You've always been a valuable member of our team, and it's going to be quite a challenge to **fill your shoes**.[6]

Your consideration for your colleagues, your professionalism, and your **determination**[7] have always impressed us. Those qualities will serve you well in your future **pursuits**.[8] It's difficult to see you go, but we're sure you'll have a lot of success in your new job.

Let us know if you're ever back in town. It'd be great to get the chance to have lunch and catch up sometime **down the road**.[9]

Best of luck,

Don Strong *Shelly Hoffman* *Mary Ann*
Ben Maples *Tony Hughes* *Scott Simmons*
Todd Andrews *Ted Parker* *Rita Duffy*

中譯

珍重再見，一路順風
黃傑克

Maples 公司十分感謝你過去十二年來的努力工作和奉獻。你一直是我們團隊中很有價值的一員，要找人接替你的職務將會是一大挑戰。

你對同事的體貼、專業精神和果斷向來讓我們留下深刻的印象，那些特質會在你將來的事業上帶來很大的幫助。看著你離職是一件困難的事，但是我們深信你的新工作一定會非常成功。

如果你有回來這裡的話，要讓我們知道。未來有機會一起吃個午餐、聊一聊近況會很棒的。

祝你好運
唐·斯壯　　雪莉·霍夫曼　瑪麗·安
班·梅普爾斯　東尼·休斯　史考特·希蒙斯
陶德·安德魯斯　泰德·帕克　瑞塔·達菲

---

7. determination [dɪˌtɜmə'neʃən] *n.* 果斷；決心

She displayed great determination in overcoming her disability.
她在克服肢體不便上展現了很大的決心。

8. pursuit [pə'sut] *n.* 職業；事務

The sales experience Sue gained will help her in future pursuits.
蘇獲得的銷售經驗將對她未來的事業有所幫助。

9. down the road 在未來

My goal is to become a CEO somewhere down the road.
我的目標是未來能成為一位執行長。

Language Corner

1. Giving Thanks for Someone's Help

I wouldn't be where I am today without all of your help.

You Might also Say . . .

- If it weren't for your help, I wouldn't have achieved what I did.
- I would never have been this successful without your assistance.

2. Keeping in Touch with Someone

It'd be great to get the chance to catch up sometime down the road.

You Might also Say . . .

- I'd love to catch up on old times in the near future.
- I look forward to maintaining our friendship for years to come.

Unit Seven
Condolences 弔唁

弔唁信函要儘快寄給死者家屬，儘量以親自手寫避免電腦打字，且長度不超過一頁。文章主要是撫慰死者家屬的傷痛，並表達關心及協助之意，內容要平實，勿用花俏用語或引經據典。需注意不宜觸及死者的死因，但可追憶其生前的生活點滴。

✔ Learning Goals

學習撰寫弔唁信函

Before We Start 實戰商務寫作句型

Topic: Expressing Consolation 表達慰問

- ◆ We've just heard about the passing / departing / loss / death of . . .
  我們剛聽到有關……過世的消息。

- ◆ Please accept our / We wish to express our / You have our deepest / sincerest / most heartfelt [condolences / sympathies . . .] on / over / about . . .
  請接受我們 / 我們想表達對……最深的 / 最真誠的 [慰問 / 同情……]。

- ◆ We were very sorry to learn of / hear about the [tragedy / catastrophe / misfortune . . .] that has befallen you / occurred.
  我們非常遺憾得知所發生（在你身上）的 [悲劇 / 災難 / 不幸……]。

- ◆ Please know / understand / remember that everyone's thoughts / prayers are with you.
  請明白 / 記得每個人都會為你祝福。

- ◆ We are all with you during this time of grief / sorrow / mourning / sadness / bereavement.
  在這個悲傷 / 痛失至親的時刻，我們全都支持你。

- ◆ (Someone) was a(n) [inspiration / trendsetter / model worker / role model . . .].
  （某人）過去是 [激勵大家的人 / 領導流行者 / 模範員工 / 典範……]。

- ◆ (Someone) was [well liked / highly respected / revered / admired / cherished . . .] by . . .
  （某人）為……所 [喜愛 / 高度尊敬 / 欽佩 / 珍愛……]。

- ◆ (Someone's) [endless contributions to the company / unmatched work ethic . . .] will sorely / greatly be missed.
  （某人）[對公司的無盡貢獻 / 無與倫比的工作道德……] 會被深深地緬懷。

- ◆ May you find [relief / solace / comfort / peace . . .] in + V-ing . . .
  願你在……中找到 [慰藉 / 平靜……]。

A. To a Grieving Client 向客戶致哀

TO: The Staff
FROM: Richard
DATE: 7/11
SUBJECT: Mr. Hanazawa's Mother
ATTACHMENT: **Condolence**[1] Letter

In case you haven't heard, Mr. Hanazawa's mother passed away over the weekend. I sent him a condolence letter (as attached) from all of Maples Corporation in an attempt to give him a small measure of support during this difficult time.

If you choose to offer your own expression of **compassion**,[2] make sure it is sent **attention to**[3] Mr. Hanazawa. Please make certain that he has our deepest **sympathies**[4] in this time of sorrow and **mourning**.[5]

It's my goal to treat all of Maples' business partners like family, especially during times like these.

All the best,
Richard

Attachment

Mr. Hanazawa
108 Sobu Street, Section 4
Japan

Mr. Hanazawa,

We've just heard about the passing of your mother. Please accept our most heartfelt condolences. We were all very sorry to learn of the tragedy that has **befallen**[6] you and wanted to express our sympathies for your loss. There are few words that one can offer to **comfort**[7] someone in a time like this, but please know that here, everyone's thoughts and **prayers**[8] are with you.

We cannot even begin to understand your feelings over losing your mother; just remember we are suffering together.

We are all with you during this time of grief.

Your friends at Maples Corporation

中譯

收件人：全體員工

寄件人：理查

日期：七月十一號

主旨：花澤先生的母親

附件：弔唁信

以防你們還沒聽說：「花澤先生的母親在週末過世了」。我以 Maples 公司全體員工的名義寄了一封弔唁信（如附件）給他，以在這個艱難的時刻給他些許支持。

如果你決定要自行表達同情，要確定信件是直接寄給花澤先生。在這個悲傷的時刻請務必向他致上我們對他最深的慰問。

我的目標是要對待 Maples 公司的生意夥伴如同家人一般，特別是在這樣的時刻。

祝 一切安好

理查

 附件

花澤先生

日本總武街四段 108 號

花澤先生：

　　我們剛聽到有關您母親過世的消息。請接受我們最真誠的哀悼。我們非常遺憾得知您所發生的悲劇，並期望能對您的喪親表達我們的慰問。在此刻沒有什麼言語能用來安慰人，但請您明白這裡的每個人都掛念著您並為您祈禱。

　　我們無法理解您的喪母之痛；只請您記住我們都感到很難過。

　　在這個悲傷時刻，我們全都支持您。

Maples 公司的朋友們 敬上

Vocabulary and Phrases

1. **condolence** [kənˋdoləns] *n.* 哀悼；弔辭（常用複數）
   Please accept my condolences for your loss.
   請接受我對你喪親的哀悼。

2. **compassion** [kəmˋpæʃən] *n.* 憐憫；同情
   He showed compassion by helping the old lady after she fell on the sidewalk.
   老太太在人行道跌倒之後，他幫了她的忙以表達憐憫之意。

3. **attention to** （寄）給
   Resumes should be sent attention to the human resources department.
   履歷表應寄到人力資源部門。

4. **sympathy** [ˋsɪmpəθɪ] *n.* 慰問；弔唁
   We offer our deepest sympathies over her death.
   我們為她的逝世獻上最深的慰問。

5. **mourning** [ˋmɔrnɪŋ] *n.* 悲傷；服喪
   The mourning of a loved one is a difficult experience for anyone.
   哀悼摯愛對任何人而言都是難熬的經歷。

6. **befall** [bɪˋfɔl] *v.* （尤指不幸）發生；降臨
   I'm sorry to learn of the tragedy that has befallen your family.
   我很遺憾得知你們家發生的悲劇。

7. **comfort** [ˋkʌmfɚt] *v.* 慰問；安慰
   The boy's mother comforted him after his baseball team lost the game.
   在那個男孩子的棒球隊輸掉比賽之後，他媽媽安慰了他。

8. **prayer** [ˋprɛr] *n.* 祈禱；禱告
   You and your family will be in our prayers.
   我們為你和你的家人祈禱。

Language Corner

Comforting Someone

There are few words that one can offer to comfort someone in a time like this.

You Might also Say . . .

- Words seem inadequate to express the sadness / sorrow we feel.
- It is difficult to say something that would make mourning any easier.
- Nothing can be said during this time of grief.

B. To a Deceased Worker's Family 向已故的同事家屬致哀

○ ○ ○

Mrs. Johnson
182 Main Street
New York, USA

Dear Mrs. Johnson,

I was very sorry to hear about the passing of your husband, Fred. All of us from the company want to convey our deepest support for you and your family. Fred was an invaluable member of our team here at Vanguard Technology. He was well liked by all the **executives**[1] and coworkers. His **endless**[2] contributions to the success and development of this company will **sorely**[3] be missed. He was truly an **inspiration**[4] to us all.

In addition to offering our sympathies for your loss, we will provide a **bereavement allowance**[5] to you and your family in this time of sorrow. We hope this will in some way ease any financial issues that may arise.

May you find comfort in knowing we all share in your sadness at this time.

Yours sincerely,
Chris Michaels
General Manager
Vanguard Technology

中譯

強森太太
美國紐約緬因街 182 號

親愛的強森太太：

　　我很遺憾聽到妳先生佛瑞德過世的消息。我們公司的全體同仁要向妳和妳的家人表達我們最大的支持。過去在 Vanguard 科技公司，佛瑞德是我們團隊中相當重要的一員。他為主管們和同事們所喜愛。他對公司成功和發展的無盡貢獻將會被深深地緬懷。他真的給予大家很多激勵。

　　除了對妳的喪夫表示哀悼之外，我們將在此悲傷的時刻提供妳和妳的家人一份喪親津貼。希望這可以多少減輕任何可能出現的財務困難。

　　希望妳能明白此刻我們和妳一起感到哀傷，願妳能藉此得到慰藉。

謹上
克里斯・麥克思
總經理
Vanguard 科技

Vocabulary and Phrases

1. **executive** [ɪgˋzɛkjʊtɪv] *n.* 幹部；管理階層

Any changes to the budget must be approved by the company's executives.
任何預算的變更都需得到公司主管的批准。

2. **endless** [ˋɛndlɪs] *adj.* 無盡的；不斷的

She has an endless supply of pens at her desk.
她的桌上有用不完的筆。

3. **sorely** [ˋsɔrlɪ] *adv.* 很；非常

The old, smelly kitchen was sorely in need of cleaning.
那個老舊難聞的廚房極需打掃一番。

4. **inspiration** [ˌɪnspəˋreʃən] *n.*
鼓舞人心的人（事物）；激勵作用

The baseball player was an inspiration to young athletes.
那位棒球選手對年輕運動員來說具有激勵的作用。

5. **bereavement allowance** 喪親津貼

We will provide NT$5,000 in bereavement allowance.
我們將提供新台幣五千元的喪親津貼。

C. Drafting a Condolence Letter 草擬哀悼信

To: Deborah Hamilton
From: Jerry Lopez
Date: May 20
Subject: Tom Hooper's Loss

Deborah,

I have just found out that Tom Hooper's father **perished**[6] in a fire. It's just unbelievable. I thought it would be nice if we could send a short letter to him expressing our sympathies. See my ideas for the letter below:

Dear Tom,

*We at Trinity Developers have just learned that you have suffered a great loss, and would like to express to you and the rest of your family our deepest sympathies and **regrets**[7] for the death of your **beloved**[8] father.*

May the love of those around you help you heal as time goes by.

(Have everyone sign)

Feel free to add anything you would like to the above sample letter.

Jerry

中譯

收件人：黛博拉‧漢彌爾頓
寄件人：傑瑞‧羅培茲
日期：五月二十號
主旨：湯姆‧胡波的喪親

黛博拉：

我剛得知湯姆‧胡波的父親在一場火災中喪生了。真令人難以置信。我想如果我們可以傳一封短信給他表示慰問，將會是很貼心的一件事。下列為我對信件的想法：

親愛的湯姆：

Trinity Developers 的同仁們剛得知你遭逢喪親，對於你摯愛的父親辭世，我們想向你和你的家人表達我們最深的哀悼和遺憾。

願你身邊的人給你的愛能幫助你隨時間撫平傷痛。

（請大家簽名）

請儘量在上面的信件範本中加上妳想說的話。

傑瑞

6. **perish** [ˋpɛrɪʃ] *v.* 死亡；喪生
I'm sorry to inform you that your uncle perished in a car accident last night.
我很遺憾要通知你，你的叔叔昨晚在一場車禍中喪生了。

7. **regret** [rɪˋgrɛt] *n.* 遺憾；哀悼
It is with heartfelt regrets that I offer my condolences to you.
我衷心遺憾地向你表達我的慰問。

8. **beloved** [bɪˋlʌvd] *adj.* 摯愛的
His beloved friend from college is coming to visit him this weekend.
這週末他大學的摯友將來探望他。

Language Corner

Presenting a Bereavement Payment

We hope the bereavement allowance will in some way ease any financial issues that may arise.

You Might also Say . . .

- We hope the bereavement payment will help with any financial problems.
- It's our hope that the bereavement allowance helps relieve any financial burden.
- We are offering a bereavement payment in the hopes that it helps keep your finances in order.

Unit Eight
Expressing Concern 表達慰問

親朋好友假如碰到天災人禍或生病、受傷、失業等不愉快的事情時，都可以寫慰問信函來向對方致意。正式的信件會以黑色或藍色筆親自手寫，但現在也有許多人為求方便而改用 e-mail。

✅ Learning Goals

學習撰寫慰問信函

Before We Start 實戰商務寫作句型

Topic: Showing Sympathy 表示同情

- We just heard about the [earthquake / typhoon / tsunami . . .] that happened in / struck / hit (place).
  我們剛聽到有關（某地）發生 [地震 / 颱風 / 海嘯……] 的消息。

- Should you require any [supplies / guidance . . .], please let us know and we will do our best to [come to your aid / be of assistance / provide support . . .].
  如果你需要任何 [補給品 / 指導……]，請讓我們知道，我們會盡力 [協助（你）/ 提供支援……]。

- All our hearts go out to you. / We're all pulling for you.
  我們都會為你祝福。

- We wish to send / offer you our [love / best wishes / encouragement . . .].
  我們要獻上我們對你的 [關愛 / 祝福 / 鼓勵……]。

- All of us are [stunned / saddened / distressed . . .] by what has befallen you / happened (to you).
  我們大家都為（你）所發生的事而感到 [震驚 / 悲傷……]。

- We are deeply concerned about your [wife / condition / diagnosis . . .].
  我們十分擔心你的 [妻子 / 狀況 / 診斷結果……]。

- . . . is an absolute [catastrophe / disaster / tragedy . . .], but I know you're going to pull through / get over this / come out in one piece.
  ……是一場十足的 [（大）災難 / 悲劇……]，但我知道你將會安然度過的。

- I know everything's going to turn out alright / fine / well / great.
  我知道一切都會好轉的。

- Best wishes for a fast / rapid / swift recovery go out to (someone).
  我衷心地期盼（某人）能快點康復。

A. Earthquake 地震

To: Jack Huang
From: Ken Smith
Date: Thursday, September 23
Subject: Is Everything OK?

Dear Jack,

We just heard the terrible news about the earthquake that **struck**[1] near Taichung! First of all, is everything OK? We hope that everyone is **safe and sound**.[2] **Needless to say**,[3] we are all quite concerned.

How's your home, Jack? Was there any structural damage? We understand about the difficulties caused by a natural **disaster**[4] of this **magnitude**.[5] Should you require any supplies, financial assistance, or **you name it**,[6] let us know and we will do our best to come to your aid.

Please take care of yourself and your family. This should be of the utmost importance.

Best wishes,
Ken

中譯

收件人：黃傑克
寄件人：肯・史密斯
日期：九月二十三號星期四
主旨：一切都好嗎？

親愛的傑克：

我們剛聽到有關台中附近發生地震的不幸消息！首先，一切都好嗎？我們希望每個人都安然無恙。不用說，我們全都相當擔心。

傑克，你的房子還好嗎？有沒有任何結構上的損壞？我們了解這種震級的自然災害所造成的困境。如果你需要任何補給品、金錢資助，或任何你可以想到的東西，要讓我們知道，我們會盡力協助你。

請好好照顧自己和你的家人。這是最重要的。

謹上
肯

Vocabulary and Phrases

1. strike [straɪk] *v.* 突然發生；侵襲

The earthquake struck off the coast of Hualien.
花蓮外海發生了地震。

2. safe and sound 安然無恙

We managed to get out of the burning building. Luckily, everyone is safe and sound.
我們設法逃出那棟起火的大樓。幸好所有人都安然無恙。

3. needless to say 當然；不用說

Our son is a famous baseball player; needless to say, we're proud of him.
我們的兒子是一位有名的棒球選手；不用說，我們以他為榮。

4. disaster [dɪˈzæstɚ] *n.* 災難

Earthquakes and fires are two common natural disasters in California.
在加州，地震和火災是兩個常見的自然災害。

5. magnitude [ˈmægnəˌtjud] *n.* 震級；強度

The earthquake had a magnitude of 7.3 on the Richter scale.
那個地震的強度為芮氏規模七點三級。

6. you name it 你儘量說

If you need any help answering the phones, handing out information, or you name it, let me know.
如果你需要幫忙接電話、分發資訊或其他任何事，要讓我知道

B. Surgery 手術

TO: Patty Smythe
FROM: Bernie Calder
DATE: November 27
SUBJECT: Get Well Soon

Dear Patty,

We wish you the best of luck with your **impending**[7] surgery. We want you to know that all our hearts go out to you. Don't forget that you are a very capable person and will undoubtedly have the **willpower**[8] to make a complete **recovery**.[9] Everyone in the office wishes to send you their love and support. We will miss you here and hope to see you soon.

Here's to a speedy return.

Your friends,
Bernie, Bobby, Gordon, and the gang

中譯

收件人：佩蒂‧史邁司
寄件人：伯尼‧卡德爾
日期：十一月二十七號
主旨：早日康復

親愛的佩蒂：

祝妳即將要動的手術能夠順利。我們想讓妳知道我們都會為妳祝福。別忘了妳是一個非常有能力的人，妳一定有讓自己完全康復的意志力。辦公室的每個人都要獻上他們對妳的關愛與支持。我們會想念妳，希望很快能見到妳。

祝妳早日歸隊。

妳的朋友們
伯尼、巴比、高登和全體同仁 敬上

---

7. **impending** [ɪmˋpɛndɪŋ] *adj.* 即將發生的；逼近的

The impending oil shortage will affect this company.
即將發生的石油短缺將影響這間公司。

8. **willpower** [ˋwɪl.pauɚ] *n.* 意志力；毅力

Jack doesn't have the willpower to finish what he has started.
傑克缺乏有始有終的毅力。

9. **recovery** [rɪˋkʌvərɪ] *n.* 復原；痊癒

Whether or not David makes a complete recovery depends on his attitude.
大衛能否完全康復取決於他的態度。

Language Corner

1. Checking On Someone

✦ How are you and your family doing?
✦ What's the latest on your son's broken leg?
✦ Could you give me an update on your grandma's condition?
✦ Fill me in on how she's doing once she leaves the hospital.

2. Making a Fast Recovery

Get well soon.

You Might also Say . . .

▪ We all wish you a quick recovery.
▪ You'll be back on your feet in no time.
▪ We hope you recuperate soon.

C. Car Accident 車禍

Dear Rolando,

We just got the news about your car accident. All of us here in the Housing Bureau are simply stunned by what has befallen you.

Most of all, we are deeply concerned about Mrs. Gonzalez, who we hear is still **in critical condition**.[2] This car accident is an absolute **catastrophe**,[3] but we know she's going to **pull through**.[4] We've been informed that she has **made big strides**[5] over the last three days. That is really good news.

To tell you the truth, this has been a very difficult letter to write. We can't imagine how **unspeakably**[6] difficult this must be for you and your family. We understand that you also suffered **minor**[7] injuries. Please try to get some rest. And don't forget to keep your chin up — we know everything's going to turn out fine. Our prayers are with you and, once again, our wishes for a swift recovery go out to your dear wife.

Yours truly,
Kathy, Stephen, Ken, Deb, and Roy

中譯 收件人：羅蘭多・龔薩雷斯
寄件人：羅伊・喬瑟夫
日期：九月四號
主旨：別氣餒

親愛的羅蘭多：

我們剛得知您出車禍的消息。房屋局的全體同仁都為您所發生的事感到震驚。

最重要的是我們十分擔心龔薩雷斯太太，聽說她仍處於危險期。這次的車禍真的是一場十足的大災難，但我們知道她將會安然度過的。我們被告知夫人在過去三天有很大的進展。那是非常好的消息。

老實說，這是一封很難下筆的信。我們無法想像這對您和您的家人來說會是多麼難以形容的煎熬。我們獲悉您也受到輕傷。請試著休息一下，並記得別氣餒。我們知道一切都會好轉的。我們都為您祈禱；再次衷心期盼您親愛的夫人能快點康復。

謹啟
凱西、史帝文、肯、黛博和羅伊

Vocabulary and Phrases

1. keep one's chin up 別氣餒；保持樂觀

Even though Lucy's husband left her, she is still keeping her chin up.
即使露西的先生離開她了，她依舊保持樂觀。

2. in critical condition 處於危險期；有生命危險

After the operation, she remained in critical condition.
手術過後，她仍處於危險期。

3. catastrophe [kə`tæstrəfɪ] *n.* 大災難

Failing one test is not a catastrophe.
一次考試不及格不是什麼大災難。

4. pull through 存活下來；度過難關

I just visited my father in the hospital. Unfortunately, he didn't pull through.
我剛去醫院探望我父親。不幸地，他沒能活下來。

5. make big strides 有很大進展

With the doctor's help, Wendy made big strides to overcome her injury.
在醫生的幫助下，溫蒂的傷勢大有進步。

6. unspeakably [ʌn`spikəblɪ] *adv.*
難以形容地；不能以言語表達地

During the earthquake, Berle was unspeakably calm.
發生地震時，伯利難以形容的鎮靜。

7. minor [`maɪnɚ] *adj.* 不嚴重的；無生命危險的

This is nothing to worry about. It's just a minor cut.
這沒有什麼好擔心的，只是輕微的割傷。

Language Corner

Encouraging Someone

Don't forget to keep your chin up — we know everything's going to turn out fine.

You Might also Say . . .

- Remember to stay positive — we know it's all going to work out in the end.
- Be sure to believe in yourself — we're confident that things will be back to normal soon.
- Keep a stiff upper lip — you'll come through this alright.

Chapter Four

Sourcing / Quotations

Unit One
Vendor Sourcing & Survey
尋找及訪視供應商

要尋找零組件供應商或 **OEM** 製造廠商,都要先派員訪查未來有可能長期合作的對象。訪查人員在參觀工廠、了解人員素質、產品品質、價格、交期、產能及配合條件等後,將訪查結果寫成報告,以供主管評估。

✅ Learning Goals

1. 學習撰寫出差工作報告
2. 了解篩選供應商所需考慮的條件
3. 了解委託代工的配合方式

Before We Start 實戰商務寫作句型

Topic: Evaluating Your Business Partner 評估生意夥伴

- ◆ I (don't) think that we should [pursue / go after / develop / establish / form . . .] this relationship / partnership / alliance.
我認為我們（不）應該 [尋求 / 發展 / 建立⋯⋯] 這段（合作）關係 / 結盟。

- ◆ The factory does not have experience / a background / expertise [in consumer products / as an OEM supplier / in the area of quality control . . .].
那間工廠沒有 [製造消費性產品 / 當代工供應商 / 品管方面⋯⋯] 的經驗 / 專業知識。

- ◆ In my mind (opinion / view) / As I see it / As it seems to me, it's [unwise / foolish / ill-advised / worthless . . .] to put full trust in / rely (depend) on the company.
就我看來，完全信賴 / 依靠那間公司是 [不智的 / 沒有用的⋯⋯]。

- ◆ The company [is clueless about / doesn't have the faintest idea about / has a total lack of experience with / is completely inexperienced with . . .] . . .
這間公司對⋯⋯ [一無所知 / 完全沒有經驗⋯⋯]。

- ◆ It seems (appears) that their prices / charges / fees were higher than other [bids / offers / proposals . . .] for our business.
他們的價錢 / 費用似乎比其他要競爭我們生意的 [出價 / 提議⋯⋯] 還要高。

- ◆ As of today / Up until now / As it stands now / At this point, we have not [made a firm commitment to / finalized an agreement with / struck a deal with . . .] any other company.
截至今天為止 / 目前，我們還沒和其他家公司 [作出確定的承諾 / 達成（最後）協議⋯⋯]。

A. Vendor Survey Report 供應商視察報告

Maples 要生產一樣新產品，但不想由原先的代工廠 Mica 和 GO 接手，於是派負責該產品開發的 Scott 帶團訪查新工廠。下列為隨行的 Richard 所撰寫的工作報告。

To: Scott Simmons
From: Richard Barlow
Date: 6/11
Subject: Trip Report
Attachment: GHMW Visit

Dear Scott,

Here is the trip report for your **perusal**,[1] as requested.

Regards,
Richard

Attachment

GUANGZHOU HEAVY MACHINERY WORKS (GHMW)
GUANGZHOU, CANTON, PRC

By Richard Barlow
June 11

SUMMARY:

On June 6, Scott Simmons, Jack Huang, and I visited the company in Guangzhou. Mr. Liang, the director of GHMW business affairs, was our host. We did not see the factory because it was closed for power **rationing**,[2] but we did see their showroom, which had **an array of**[3] auto **replacement parts**.[4]

During a post-tour meeting, Mr. Liang asked about scheduling as well as the test equipment and **fixtures**[5] needed. We agreed to send an equipment list and a line layout.* They were to tell us which parts they could source themselves.

Mr. Liang was concerned about the technical support we would provide, since they had no experience with our products. We said we'd supply two on-site engineers. Then, Mr. Liang asked about a deposit that he thought we should pay **up front**[6] for production preparations — a request that is unheard of for an OEM relationship. As we grew to understand what he was saying, it became clear that they had never done an OEM job before.

COMMENT:

I don't think that we should pursue this relationship. The factory does not have experience in consumer products or as an OEM supplier. In my mind, it's unwise to put full trust in a company that is brand-new to the process. They do not have any engineers who speak English. Furthermore, they are **clueless**[7] about requirements and procedures for clean rooms.* They would require **massive**[8] support from us to get a line going.

With the question of the up-front payment and cost of supporting a factory in Guangzhou, the quote doesn't look very good anyway.

中譯

收件人：史考特‧希蒙斯
寄件人：理查‧巴洛
日期：六月十一號
主旨：出差報告
附件：參觀廣州重機工廠

親愛的史考特：

如您所要求的，附上這次的出差報告供您參詳。

祝 商安
理查

 附件

廣州重機工廠
中國廣東廣州

理查‧巴洛　撰

六月十一號

摘要：

　　本人和史考特‧希蒙斯與黃傑克於六月六號參訪了這間在廣州的公司。由廣州重機工廠的營業部經理梁先生接待我們。工廠因分區輪流供電的緣故而關閉，所以我們無法參觀。但我們有參觀他們的展示間，那裡有一系列的汽車備用零件。

　　在參訪後的會議中，梁先生問到排程和所需的測試設備及固定裝置。我們同意寄給他們一份設備清單和生產線規劃圖，而他們會告訴我們哪些零件他們可以自行採購。

　　梁先生很關心我們在技術方面所提供的支援，因為他們沒有製造我們產品的經驗。我們表示會派兩位工程師到現場。之後梁先生問到有關訂金的事，他以為我們會預付準備生產的費用（在合作代工方面未曾聽過這種要求）。在逐漸明白他所說的話後，我們可以很清楚的知道他們從未做過代工的工作。

評論：

　　我認為我們不應該尋求這段合作關係。那間工廠沒有製造消費性產品或當代工供應商的經驗。就我看來，完全信賴一間對製程毫無了解的公司是不智的。他們沒有工程師會說英文，對無塵室的規定和程序也一無所知。他們將會需要我們大量的支援才能進行生產。

　　預付費用的問題再加上支援廣州工廠的成本，總之他們的報價看起來不太理想。

Vocabulary and Phrases

1. **perusal** [pə`ruzl] *n.* 細讀；仔細研究（peruse *v.*）
   I've attached a copy of the contract for your perusal.
   我附上一份合約的副本供你參詳。

2. **rationing** [`ræʃənɪŋ] *n.* 定量配給
   The troops were low on food, so the general had to do some food rationing.
   部隊的食物不足，所以將軍必需實施食物定量配給。

3. **an array of** 大量；一批；一系列
   The museum has an array of items from the 1700s.
   那間博物館有一系列十八世紀的館藏。

4. **replacement part** 備用零件（= spare part）
   My car needed to be fixed, so I went to the store to look for replacement parts.
   我的車需要修理，所以我到那家店找備用零件。

5. **fixture** [`fɪkstʃɚ] *n.* 固定裝置；設備；（工件）夾具
   We used fixtures to keep the parts steady during the manufacturing process.
   我們在製造過程中使用夾具來固定零件。

6. **up front** *adv.* 預先（up-front *adj.*）
   I didn't want to pay up front for the new motorcycle.
   我不想預先支付那台新摩托車的費用。

7. **clueless** [`klulɪs] *adj.* 一無所知的；無線索的
   He's totally clueless about running a company.
   他對於經營一間公司毫無頭緒。

8. **massive** [`mæsɪv] *adj.* 大量的；大規模的
   The damage done by the typhoon will require a massive recovery effort.
   這個颱風所造成的傷害將需要大規模的恢復重建工作。

Biz Focus

★ line layout　生產線規劃圖或布置圖　　★ clean room　無塵室

Language Corner

Word Usage — Grow

grow + to V. / adj.　漸漸變成……

For example:

1. Stacy didn't like her roommate at first, but she grew to like her.

2. Daniel grew tired of hearing his coworker's complaints.

B. Looking for Suppliers 尋找供應商

Maples 老闆 Ben 的朋友 Joseph 要在台灣找配合的供應商，於是 Ben 請台灣分公司的 Jack 幫忙安排。

To: Jack Huang
From: Joseph Perry
Date: February 25
Subject: Our Previous Meeting

Dear Jack,

I finally returned home last week. This is the first chance I've had to write thank-you notes and to catch up on **correspondence**.[1]

I would like to **impart**[2] to you how much I enjoyed meeting you and the three gentlemen from Morrison Precision Industrial. Just to give you an update about our thought process, it seems that their prices were higher than other **bids**[3] for our business. As of today, we have not made a **firm**[4] commitment to any other company. Nevertheless, I doubt if we can afford their services at this time.

I will be seeing Ben in Los Angeles within the month, and I will **make a point of**[5] telling him how you tried to help find qualified suppliers.

I hope the next time I come to Taipei, we can **rendezvous**[6] again. I'd like to spend a little more time reviewing other opportunities.

Kindest regards,
Joseph Perry

中譯 收件人：黃傑克
寄件人：喬瑟夫・派瑞
日期：二月二十五號
主旨：我們之前的會面

親愛的傑克：

　　我終於在上週回到家了。這是我第一次有機會可以寫感謝信並查看我的信件。

　　我想讓你知道我非常高興能與你和 Morrison Precision Industrial 的三位先生見面。也讓你知道一下我們現在的想法，他們的價錢似乎比其他要競爭我們生意的出價還要高。截至今天為止，我們還沒和其他家公司作出確定的承諾。不過，這次我們恐怕負擔不起雇用他們的費用。

　　我這個月會到洛杉磯拜訪班。我會向他特別強調你是如何盡力地幫我們尋找合適的供應商。

　　希望我下次到台北時，我們可以再次碰面。我想再花點時間討論其他合作機會。

　　謹上
　　喬瑟夫・派瑞

Vocabulary and Phrases

1. **correspondence** [ˌkɔrəˈspandəns] *n.*
（總稱）信件；通信聯繫

I don't have Internet access, so I can't keep up with my correspondence.
我沒有網路，所以我無法保持通信聯繫。

2. **impart** [ɪmˈpɑrt] *v.* 告知；透露

The lawyer tried to impart to the jury that his client was innocent.
那名律師試圖讓陪審團了解他的當事人是無辜的。

3. **bid** [bɪd] *n.* 出價；投標

The wealthy businessman placed a bid on the sports car at the auction.
那位富商在拍賣會上出價競標那台跑車。

4. **firm** [fɜm] *adj.* 確定的；不會改變的

Susan hasn't made any firm plans for the weekend.
蘇珊這週末沒有確定的計畫。

5. **make a point of** 特別努力做某事

The manager made a point of giving credit to his staff for the store's success.
因為店舖的成功，那名經理特地讚揚他的員工們。

6. **rendezvous** [ˈrɑndəˌvu] *v.* 約會；會面

We'll rendezvous with our friends at four p.m. in the hotel lobby.
我們今天下午四點要和我們的朋友們在飯店大廳碰面。

Language Corner

Showing Appreciation

I will make a point of telling him how you tried to help find qualified suppliers.

You Might also Say . . .

- I will emphasize to him your efforts to find suitable suppliers.
- I will make a special effort to tell him how you attempted to locate suppliers that can serve our needs.
- I will go out of my way to let him know how you tried to help source capable suppliers.

Unit Two
Component /Assembly
Sourcing 採購零組件

欲採購零組件時，需先以傳真或電子郵件大致確認廠商是否有生產所需的產品及可能的報價，之後再郵寄詳細圖面及規格，甚至提供實際樣品請廠商正式報價。

 Learning Goals

1. 買主要儘量避免介入 OEM 工廠零件的採購工作

2. 詳細產品資料，如用途、品質要求等，可協助廠商確實報價

3. 採購零組件除考量價格外，也需評估交貨期、廠商是否能直接出貨報關等條件

Before We Start 實戰商務寫作句型

Topic 1: Introducing General Information 提供資訊

- For your background information, . . .
- To give you a general summary, . . .
- To provide you with a broad picture of . . .
- Just so you have a comprehensive idea of . . .
  讓你大略了解一下，……

Topic 2: Handling Sourcing Issues 處理採購事宜

- In the spirit of cooperation, we sourced / located / identified a few / a couple (of) / several [vendors / buyers / distributors . . .] in (place) for . . .
  本著合作的精神，我們在（某地）替……找了一些 [供應商 / 買家 / 批發商……]。

- The company is interested in / has (shows / expresses) interest in / wants to find out more about [ship components / electronic parts / raw materials / computer chips . . .].
  那間公司對 [船隻零件 / 電子零件 / 原料 / 電腦晶片……] 有興趣 / 想進一步了解。

- They've swamped / flooded / overwhelmed / inundated / deluged me with [drawings / samples / designs . . .] looking for / asking for / requesting quotes for . . .
  他們給我非常多 [圖面 / 樣品 / 設計……] 要……的報價。

- If I sent to you a bundle / batch / bunch / set of drawings, could you reciprocate with / reply with / get back to me with [part costs / terms / delivery schedules . . .]?
  如果我寄給你一組圖面，你可以回寄給我 [零件成本 / 條件 / 送貨時間表……] 嗎？

Richard 寫信給代工廠 GO 的品管經理，說明 MAT 只幫忙 GO 找零件供應商，如所採購的零件品質有問題，應直接找供應商負責。

○ ○ ○

To: Alfred Kerts
From: Richard Barlow
Date: 9/4
Subject: Component Sourcing

Dear Alfred,

Yesterday, we received a box of **defective**[2] parts from GO. They were returned to us because they were **allegedly**[3] rejected by your IQC* people as having out-of-spec* dimensions. For some reason, your people think they can send us their production-line defects and have us convince *your* vendor to replace them. I am returning the pieces to you so they'll be at your **disposal**.[4]

For your background information, there are several components that the GO purchasing people could not locate, so in the spirit of cooperation, we sourced a few vendors in Taiwan. I'm sorry my group got involved in sourcing components for GO. Every vendor we introduced to GO has come back to us complaining about **nonpayment**[5] of invoices, **unresponsiveness**[6] to inquiries, and the returning of defective components without QC reports or explanations of the problem. I know you can't do anything about the nonpayment and communication issues, but I hope you'll take immediate action to correct the **abysmal**[7] QC situation.

Best regards,
eom

中譯　收件人：艾佛列‧柯茲
寄件人：理查‧巴洛
日期：九月四號
主旨：零件採購

親愛的艾佛列：

昨天我們收到一箱 GO 寄來的瑕疵零件。據説因為你們的進料品管人員認為它們的尺寸不合規格而驗退，所以將它們寄回給我們。由於某種原因，你們的員工認為他們可以把生產線上的瑕疵品寄給我們，並要我們去説服「你們的」供應商來做更換。我現在把這些零件退還給你，讓你自行處置。

讓你大略了解一下，有一些零件 GO 的採購人員找不到，因此本著合作的精神，我們在台灣找了一些供應商。我很抱歉我們公司涉入替 GO 採購零件的工作。每個我們介紹給 GO 的供應商都會回過頭來向我們抱怨有關未付款、對詢問置之不理及在未附品管報告或對問題沒有解釋的情況下退還瑕疵的零件。我知道有關未付款和溝通方面的問題你無能為力，但我希望你能立即採取行動來改善這種糟糕的品管情況。

謹上
完畢

Vocabulary and Phrases

1. **procurement** [proˈkjʊrmənt] *n.* 採購；獲得（procure *v.*）
   I'll take care of the procurement of funds from customers.
   我將會負責從客戶那裡取得資金。

2. **defective** [dɪˈfɛktɪv] *adj.* 有缺陷的；有瑕疵的（defect *n.*）
   I'm returning the computer I just bought because it is defective.
   我要退還我剛買的電腦，因為它有瑕疵。

3. **allegedly** [əˈlɛdʒɪdlɪ] *adv.* 據宣稱；據傳說
   Allegedly, the company will be sold next month.
   據說這間公司下個月將要被賣掉。

4. **disposal** [dɪˈspozl̩] *n.* 處理；處置
   The candidate has millions of dollars at his disposal.
   那位候選人有幾百萬美金任他處置。

5. **nonpayment** [nɑnˈpemənt] *n.* 未（予、拒）付
   Lydia was asked to move out of her apartment for nonpayment of rent.
   麗狄亞因為未付房租而被要求搬離她的公寓。

6. **unresponsiveness** [ˌʌnrɪˈspɑnsɪvnɪs] *n.* 沒有答覆；沒有反應
   He assumed from her unresponsiveness to his phone calls that she was asleep.
   由於她沒有接電話，他假定她已經睡著了。

7. **abysmal** [əˈbɪzml̩] *adj.* （口語）糟透的；極壞的
   The company posted abysmal profits last month.
   那間公司上個月公布的盈利非常差。

Biz Focus

★ IQC 進料（貨）品管（為 Incoming Quality Control 的簡寫）

★ out-of-spec 不在規格內的

Language Corner

How to Give a Firm Request

I hope you'll take immediate action to correct the abysmal QC situation.

You Might also Say . . .

- It would be in your interests to act at once to amend the terrible QC situation.

- I expect you'll get right to work on fixing the QC situation.

B. PCB Sourcing 採購印刷電路板

James 寫信給香港專門供應印刷電路板的廠商，請他們提供三種不同印刷電路板的報價。

To: Bob Davies
From: James Simon
Date: July 10
Subject: PCB Quotations

Dear Bob,

I'm sorry to have confused you with my **haphazard**[1] **references**[2] to a different project we have cooking. Currently, we have three PCB quotations **in the works**.[3]

The first one is for my own product called TouchFree, which is a **gizmo**[4] that is easy to install in a sink. It automatically turns the water on when it **ultrasonically**[5] senses something (hands, a plate, etc.). When the sensed object goes away, the water is turned off. We can use TF-1 as a P / N* for TouchFree. I sent you a **packet**[6] today with information about the product.

The second project I've asked you about is a bare board.* This will be used for high-voltage power supplies. At this time, the buyer is sourcing from Taiwan. I know your company doesn't produce bare boards, but I thought your industry contacts would **come in handy**,[7] and maybe we could both make some **dough**[8] on the deal.

Lastly, I have info about a PCB assembly job in the mail to me now. I don't know the details right now. I'll pass them on to you when I do.

Cheers,
James

中譯

收件人：鮑伯・戴維斯
寄件人：詹姆士・西門
日期：七月十號
主旨：印刷電路板報價

親愛的鮑伯：

很抱歉因為我隨口提到另一個我們正在進行中的案子，而讓你感到困惑。目前我們有三個印刷電路板的報價正在處理中。

第一個是我自己的產品叫做 TouchFree，它是一個可以輕易安裝在水槽的小玩意兒。當它經由超音波感應到東西（手、盤子等）時，就會自動把水打開。當被感應到的物品移開時，水就會關掉。我們可以用 TF-1 作為 TouchFree 的物品編號。我今天寄了一個小包裹給你，裡面有這個產品的資訊。

我向你詢問的第二個案子是一塊裸板。這塊板子將用於高電壓電源供應器。目前買主是從台灣採購的。我知道貴公司並不生產裸板，但我想你們在業界的人脈可以派上用場。或許我們雙方都可以在這個交易中賺一些錢。

最後，我將會收到一份有關組裝印刷電路板的資料，我現在還不曉得細節，我知道後會再告訴你。

工作愉快
詹姆士

Vocabulary and Phrases

1. **haphazard** [ˌhæpˈhæzɚd] *adj.* 隨意的；雜亂的

   The business trip was arranged in a haphazard way.

   這次的出差毫無規劃可言。

2. **reference** [ˈrɛfərəns] *n.* 提及；涉及

   The broadcaster got in trouble for her reference to Hitler on air.

   那名廣播員因為在節目中提及希特勒而惹上麻煩。

3. **in the works** 處理中；進行中

   I've got three possible sales in the works right now; I'm hoping one will work out.

   我有三個可能的銷售案正在進行中；我希望有一個能成功。

4. **gizmo** [ˈgɪzmo] *n.* （口語）新玩意兒；小發明

   （亦可拼成 gismo）

   The movie featured all sorts of high-tech gizmos.

   那部電影的特色是有各種高科技的新玩意兒。

5. **ultrasonically** [ˌʌltrəˈsɑnɪklɪ] *adv.* 利用超音波地

   The car bumper ultrasonically senses its distance from an object and notifies the driver.

   那部車子的保險桿會利用超音波感應和物體的距離，並且告知駕駛。

6. **packet** [ˈpækɪt] *n.* 小包（裹）

   The travel agent sent her a packet about hotels on the island.

   那位旅遊仲介寄給她一個有關島上飯店的小包裹。

7. **come in handy** 派上用場

   You should keep that receipt. It could come in handy.

   你應該保留那張收據。它可能會派上用場。

8. **dough** [do] *n.* （口語）錢；鈔票

   I worked a lot of hours this month. That's why I have so much dough.

   我這個月的工作時數很多，所以我有這麼多錢。

Biz Focus

★ PCB 印刷電路板（= Printed Circuit Board）

★ P / N 部品編號；零件型號（= Part Number）

★ bare board 裸板；空電路板

Language Corner

1. Looking for Opportunities

Your industry contacts would come in handy.

You Might also Say . . .

- This networking of yours could be quite useful.
- Your connections in the field might turn out to be helpful.

2. Making Money

✦ make some dough
✦ make a shekel
✦ make a fortune
✦ make some clams
✦ make some greenbacks
賺錢

Sean 在拜訪荷蘭一家採購公司並取得一些估價用的圖面後，寫信給大陸一家工廠詢問是否可由他們來提供報價。

To: Leo Kuo
From: Sean Maguire
Date: 2/27
Subject: Project Update

Hello Leo,

Thank you for your e-mail dated February 8. I apologize for not responding sooner, but I just returned from a trip through the US and Europe that was a combination of business and pleasure. It was a great trip. It's time to get back to work.

The steam-iron project is still alive, but we've done nothing **substantive**[1] on it since we talked with you. Anyway, on to the business at hand. To give you a general summary, I went to Holland to meet with a new customer, which sources components — mostly machined and cast-metal parts (all sizes) — for European companies. This is the same company that is interested in heavy-ship components (**hatches**,[2] shell doors, etc.). They've **swamped**[3] me with drawings looking for quotes for aluminum die-cast parts. I'm writing to you to find out how I should **proceed**.[4]

Since I haven't done business with companies in China, I must start by asking some **fundamental**[5] questions. If I sent to you a bundle of drawings, could you **reciprocate**[6] with tooling and part costs? How about terms and delivery schedules? I assume that you could ship directly to Europe without needing me to get involved. Is this true?

I'll look forward to receiving your **counsel**.[7]

Sincerely,
Sean

中譯　收件人：郭里歐
寄件人：尚恩・馬奎爾
日期：二月二十七號
主旨：案子的最新進展

里歐，你好：

　　謝謝你於二月八號的來信。很抱歉我沒有早一點回覆，我剛從美國和歐洲回來，去出差順便度假。那是趟非常棒的旅程，現在是時候回到工作崗位了。

　　蒸氣熨斗的案子還在進行，但是自從我們和你談過之後，就一直沒有什麼實質的作為。總之，言歸正傳。向你簡報一下，我到荷蘭和一個新客戶會面，他們主要為歐洲公司採購各種尺寸的機械和金

屬鑄造零件。同一家公司也對重型船隻零件（艙口、門殼等）很有興趣。他們給我非常多的圖面，要鋁製壓鑄零件的報價。我寫這封信給你是想詢問我應該如何進行。

　　因為我從未與大陸的公司做過生意，我必需從詢問一些非常基本的問題開始。如果我寄給你一組圖面，你可以回寄模具和零件的成本給我嗎？條件和運送時間表呢？我假定不需要我的介入你們就可以直接出貨到歐洲，對吧？

　　我期待能得到你的建議。

祝 商安
尚恩

Vocabulary and Phrases

1. **substantive** [`sʌbstəntɪv] *adj.* 實在的；實際的

   They don't have any substantive evidence to prove that he is guilty.
   他們沒有任何實際的證據可以證明他是有罪的。

2. **hatch** [hætʃ] *n.* （船的）艙口；入口

   The captain opened the hatch and climbed out of the submarine.
   那個船長把艙口打開並爬出了潛艇。

3. **swamp** [swɑmp] *v.* 使應接不暇；使陷入困境

   We've been swamped with meetings all day.
   我們一整天都不斷在開會。

4. **proceed** [prə`sid] *v.* 繼續進行；繼續做（講）下去

   The city government allowed workers to proceed with the construction of the hotel.
   市政府同意讓工人們繼續進行那間飯店的建設工程。

5. **fundamental** [ˌfʌndə`mɛntḷ] *adj.* 基礎的；根本的

   Reading and writing are fundamental skills in any language.
   閱讀和寫作是任何語言的基本技能。

6. **reciprocate** [rɪ`sɪprəˌket] *v.* 交換；互給

   I bought Ben a birthday gift, so I hope he'll reciprocate on my birthday.
   我買了生日禮物給班，所以我希望他在我生日的時候也能回送。

7. **counsel** [`kaʊnsḷ] *n.* 忠告；勸告

   Thanks for your counsel on how to make our company more efficient.
   謝謝你對如何讓我們公司更有效率所提出的忠告。

Biz Focus

★ die-cast 鑄模的

Language Corner

1. Moving to the Main Topic

Anyway, on to the business at hand.

You Might also Say . . .

- Anyway, let's focus on our current concerns.
- Anyway, let's turn our attention to the task we're facing.

2. Asking for Instructions

I'm writing to you to find out how I should proceed.

You Might also Say . . .

- Please kindly tell me what I should do next.
- I'm e-mailing to ask you to point me in the right direction.

Unit Three
Quote Inquiries
詢價

不管是經由介紹或從媒體廣告中得知某家公司所提供的產品或服務，有興趣者可以寫信給對方，在取得進一步的產品資訊及報價後，再決定是否要下單採購。

✅ Learning Goals

1. 詢問信的開頭會先表明從何管道得知對方，再說明主要訴求

2. 收到詢問信時，需先回覆確認收到信並預告何時會答覆。如無法提供所需資訊，也要委婉地解釋原因

Before We Start 實戰商務寫作句型

Topic: Talking about Quotation Specifics 談論報價細節

- After (a) careful review of / (a) thorough examination of / a closer look at your [operations / services . . .], my organization believes that your company is ideal for serving / meeting / satisfying / fulfilling our needs of . . .

  在仔細檢視你們的 [營運 / 服務……] 後，我們認為貴公司非常符合我們……的需求。

- We'd like to know further your terms and conditions as related to / pertaining to / regarding the services that we would be requiring / requesting.

  我們想要進一步了解有關我們所要求的服務，你們的條件是什麼。

- We'd require from (company) the service(s) of [sourcing and QC / manufacturing . . .] adhering to / conforming to / complying with / following the above [specifications / requirements / standards . . .].

  我們需要（某公司）遵照上述的 [規格 / 要求 / 標準……] 來提供 [採購及品管 / 生產……] 的服務。

- If you're interested in quoting this product, please [prepare / formulate / draw up . . .] a quotation based on / on the basis of / grounded on / according to the following . . .

  如果你有興趣要對該產品報價，請根據下列內容 [準備 / 制訂 / 草擬……] 一份報價單……

- I am writing (this letter) to inquire about the potential / possibility / prospect / likelihood of [securing / obtaining / signing . . .] an agent who will . . .

  我寫此封信是想詢問關於 [找到 / 簽約雇用……] 代理商來……的可能性。

- I hope to hear from you soon / to receive your prompt response / you will reply swiftly with a quote and [further / much more detailed / additional . . .] information.

  希望你能儘速回覆，提供報價及 [進一步 / 更詳細 / 額外……] 的資訊。

A. Initial Business Talks 初步商業會談

#1 有客戶在雜誌上看到 FETS 的廣告，於是與 FETS 聯絡，希望他們在台灣代其採購二手摩托車並作出貨檢驗

To: Richard Barlow
From: Jon Davis
Date: 25th March
Subject: Used Motorbikes

Dear Mr. Barlow,

During my February visit to Taipei, I spoke to Mr. Jack Huang, who later sent information on your company. After careful review of your operations and **manpower**,[1] my organization believes that your company is ideal for serving our Asian market needs. **To this effect**,[2] we'd like to know further your terms and conditions as related to the services that we would be requiring from you.

We are interested in the **importation**[3] of used motorbikes from Taiwan — no more than five years old with good bodies and perfect working engines. We'd require from FETS the services of sourcing and QC **adhering to**[4] the above specifications and **guaranteeing**[5] the bikes' condition before shipment.

We eagerly await your reply.

Yours faithfully,
Jon Davis
Managing Director
BARA ROSE LTD.

中譯

#1

收件人：理查・巴洛
寄件人：強・戴維斯
日期：三月二十五號
主旨：二手摩托車

親愛的巴洛先生：

在我二月去台北拜訪的期間，我曾與黃傑克先生談過，稍後他也將你們公司的資訊寄給我。在仔細檢視你們的營運和人力後，我們認為貴公司非常符合我們在亞洲市場的需求。為此我們想要進一步了解有關我們所要求的服務，你們的條件是什麼。

我們有意從台灣進口二手摩托車一車齡五年以下、外型完好，有完整可運作的引擎。我們需要 FETS 遵照上述的規格來提供採購及品管的服務，並在出貨前確保摩托車的狀態。

我們殷切等候您的回信。

謹上
強・戴維斯
總經理
BARA ROSE LTD.

Vocabulary and Phrases

1. manpower [ˋmænˌpauɚ] *n.* 人力；人力資源
The construction company didn't have enough manpower to finish the project on time.
這間建設公司沒有足夠的人力準時完成這個案子。

2. to this effect 為此
I wanted him to remember; to this effect, I sent him a series of e-mails.
我想讓他記得；為此，我寄給他一連串的電子郵件。

3. importation [ˌɪmporˋteʃən] *n.* 進口；進口貨
The company deals with the importation of valuable jewelry.
那間公司經營貴重珠寶的進口生意。

4. adhere to 遵照
If they don't adhere to the guidelines, they'll be punished.
如果他們沒有遵照指導方針將會被處罰。

5. guarantee [ˌgærənˋti] *v.* 保證；擔保
The auto mechanic guaranteed his work for up 90 days.
那名汽車技師保證有多達九十天的保固期。

6. considerable [kənˋsɪdərəbl̩] *adj.* 相當多（大）的
He's donated a considerable amount of mone the hospital.
他捐贈了一大筆錢給那間醫院。

#2 有關對方的需求，Richard 認為 FETS 不適合幫客戶在台灣採購二手摩托車，但同意代為作出貨檢驗。

To: Jon Davis
From: Richard Barlow
Date: April 6
Subject: Re: Used Motorbikes

Dear Mr. Davis,

Thank you for your interest in our company.

We do a **considerable**[6] amount of component and product sourcing, but we've never sourced used motorbikes. I don't think we could compete with the people you've already negotiated with. I'm of the opinion that your best **approach**[7] would be to talk directly with the suppliers.

We still stand ready to inspect and expedite the shipment of used motorbikes you buy here; however, I don't believe we are in a position to offer you service in procuring the items.

In any case, thanks again for your interest in FETS. Please keep us in mind in the future. Good luck with your **venture**.[8]

Sincerely,
Richard Barlow

中譯

#2

收件人：強‧戴維斯
寄件人：理查‧巴洛
日期：四月六號
主旨：回覆：二手摩托車

親愛的戴維斯先生：

謝謝您對我們公司有興趣。

我們採購過相當多的零件和產品，但未曾採購過二手摩托車。我想我們可能無法與您已斡旋過的其他業者競爭。我認為您最好的作法就是直接與供應商洽談。

我們仍然可以為您檢驗您在台灣購買的二手摩托車並負責催貨；然而，我想我們無法提供你們採購這些產品的服務。

無論如何，再一次感謝您對 FETS 有興趣，未來也請繼續將我們考慮在內。祝貴公司鴻圖大展。

祝 商安
理查‧巴洛

---

7. **approach** [ə`protʃ] *n.* 處理的方法或手段

His sales approach involved talking with customers about their personal interests.
他的銷售手法是和顧客們聊他們的個人興趣。

8. **venture** [`vɛntʃɚ] *n.* （帶風險的）事業；投機活動

Our latest venture is to start a cellular service company in China.
我們最新的事業是在中國開設一家手機服務公司。

Language Corner

How to Politely Reject a Customer

In any case, thanks again for your interest in FETS. Please keep us in mind in the future.

You Might also Say . . .

- No matter what happens, thank you for inquiring about our services. Don't hesitate to contact us if we can assist you later on.
- Whatever happens, we're grateful for your interest in FETS. We hope we can help with future projects.

B. Request for a Quotation 要求報價

Maples 公司以前的員工 Roy 在離職後幫客戶設計自動調溫器，他委託 Jack 替他在台灣詢價。

To: Jack Huang
From: Roy Mack
Date: January 4
Subject: **Thermostat**[1] Quote

Dear Jack,

First let me start by wishing you and your staff a **belated**[2] Merry Christmas and Happy New Year.

Enclosed please find a set of drawings for the thermostat project I discussed with you. A thermostat is a device that **regulates**[3] the temperature of the heating and / or air-conditioning system in a house so that the system's temperature is maintained near a desired set-point temperature.

If you are interested in quoting this product, please prepare a quotation based on the following:

1. An annual quantity of 150,000 to 250,000 units

2. FOB* Taiwan

3. Black-and-white **instruction booklets**[4] (not defined)

4. Four-color display boxes (not defined)

5. A function test time — three minutes

6. Artwork* and color samples supplied with the order

The quotation should be completed **prior to**[5] Chinese New Year so that we can get back to you in a timely manner. As an **aside**,[6] before the **awarding**[7] of an order, my client will want to visit you and your manufacturer. If you have any questions, please don't hesitate to contact me.

Kind regards,
Roy

中譯 收件人：黃傑克

寄件人：羅伊・馬可

日期：一月四號

主旨：自動調溫器報價

親愛的傑克：

首先獻上我遲來的祝福，祝你和你的同事們在剛過的聖誕節和新年都過得很快樂。

附件為我和你討論過的自動調溫器案子的一組圖面。自動調溫器是一個可以調節室內暖氣和

（或）冷氣溫度的裝置，這樣系統的溫度就可以維持在所希望的定點溫度上下。

如果你有興趣要對該產品報價，請根據下列內容準備一份報價單：

1. 每年數量為十五萬到二十五萬台

2. 台灣船上交貨的離岸價格

3. 黑白使用手冊（未定案）

4. 四色展示盒（未定案）

5. 功能測試時間 — 三分鐘

6. 包裝設計圖和彩色的樣本會隨訂單提供

報價要在農曆新年前完成，這樣我們才能及時回覆你。岔題一下，在下訂單之前，我的客戶想要拜訪你和你的製造商。如果你有任何問題，請盡量和我聯絡。

謹上
羅伊

Vocabulary and Phrases

1. **thermostat** [ˈθɜrməˌstæt] *n.* 自動調溫器；恆溫器

   I want to check the temperature. Where's the thermostat?
   我想查看一下溫度。恆溫器在哪裡？

2. **belated** [bɪˈletɪd] *adj.* 遲來的；延誤的

   I want to wish you a belated Happy New Year.
   我要跟你拜個晚年。

3. **regulate** [ˈrɛgjəˌlet] *v.* 調節；校準

   The car's tires have sensors to regulate air pressure.
   那部車的輪胎有感應器可以調整胎壓。

4. **instruction booklet** 使用手冊；操作說明書

   The father opened the instruction booklet so he could put together his son's new toy.
   那位父親打開操作說明書，以便組合他兒子的新玩具。

5. **prior to** 在……之前

   I have to get this sales report done prior to going on vacation.
   我必須在休假前完成這份業務報告。

6. **aside** [əˈsaɪd] *n.* 離題的話

   As an aside, I'd like to have a short meeting with you about a different matter.
   岔題一下，我想就另一件事和你開個簡短會議。

7. **award** [əˈwɔrd] *v.* 給予；授予

   After considering his experience, the supervisor awarded him the job.
   在考量他的經驗之後，主管給了他這份工作。

Biz Focus

★ **FOB** 船上交貨或離岸價格（= Free on Board）

其報價通常包含工廠價、報關費、拖車費、港口吊櫃費等。

★ **artwork** （此指）產品包裝設計圖

Language Corner

Setting a Deadline

The quotation should be completed prior to Chinese New Year so that we can get back to you in a timely manner.

You Might also Say . . .

- The quotation should be finished before Chinese New Year so we can respond quickly.

- The quotation should be finalized in advance of Chinese New Year so that we can reply in a swift fashion.

加拿大客戶第一次在台採購高爾夫球桿，因擔心出貨品質而親自來台驗貨又不划算，故委託 FETS 作品管檢驗的工作。

○ ○ ○

To: Richard Barlow
From: Stephen Cameron
Date: Jan. 9
Subject: Service Agent Needed

Dear Mr. Barlow,

I am writing this letter to inquire about the potential of **securing**[1] a service agent who will complete a final quality control* inspection. We currently have an L / C opened in favor of* a golf club manufacturer in Taichung. The L / C is not large; however, as we are just establishing a relationship with this company, we feel it is **prudent**[2] to progress at a reasonable rate.

The L / C details are as follows:

1. Invoice Total: US$20,935

2. **Particulars**:[3] 330 individual golf clubs **composed**[4] of three models

3. Inspection Time: end of February

4. Per-club Cost Ranges: from US$53 to US$68

The inspection is quite simple and can be completed in one to two hours. My main concerns would be:

1. Artwork, color, and tooling **compliance**[5]

2. Proper model numbers

3. Testing shaft lengths and club weights of approximately five to 10 **randomly**[6] selected clubs

This inspection is more visual than technical and, as it is in the Taichung area, I hope working with you will make it as **economical**[7] as possible for us. Of prime importance to our company is the permanent establishment of a relationship with a **reputable**[8] service agent who would complete our larger and more numerous inspections in the future. I hope to hear from you soon with a quote and further information.

Yours truly,
Stephen Cameron
President
Prudential Lenders

中譯

收件人：理查‧巴洛
寄件人：史帝文‧卡麥隆
日期：一月九號
主旨：徵求服務代理商

親愛的巴洛先生：

　　我寫此封信是想詢問關於找一個服務代理商來完成成品檢驗的可能性。我們目前開了一張信用狀給位在台中的高爾夫球桿製造商。信用狀的金額不大，但是因為我們剛開始與這間公司有生意往來，所以我們覺得以合理的速度來進行是明智的。

　　信用狀的細節如下：

1. 發票總金額：兩萬零九百三十五元美金
2. 細目：含三種型號的高爾夫球桿三百三十支
3. 檢驗時間：二月底
4. 球桿單價範圍：五十三元美金到六十八元美金

　　檢驗相當簡單，可在一到兩個小時之內完成。我的主要考量是：

1. 包裝設計、顏色和模具需符合要求
2. 型號正確
3. 隨機選取約五到十支球桿來檢驗其桿身的長度和球桿的重量

　　這項檢驗比較偏向外觀而非技術層面，而且就在台中地區檢驗，所以我希望和你們合作會幫我們儘可能的省錢。對我們公司而言最重要的是：能與在未來有辦法完成更大規模、更多次數檢驗，且有信譽的服務代理商建立長久的關係。希望你能儘速回覆，提供報價及進一步的資訊。

敬上
史帝文‧卡麥隆
總裁
Prudential Lenders

Vocabulary and Phrases

1. secure [sɪ`kjʊr] *v.* 獲得；替……弄到

We're trying to secure a new sales manager by the end of the month.
我們試著在月底前找到新的業務經理。

2. prudent [`prudn̩t] *adj.* 精明的；謹慎的

It would be prudent to research all the issues before you vote.
在投票前研究所有的議題是明智的。

3. particular [pə`tɪkjələ] *n.* 細目；詳細情況（作此義解時，常用複數）

For all of the particulars, you need to consult Joe.
有關所有詳情，你得請教喬。

4. compose [kəm`poz] *v.* 組成；構成

The basketball league is composed of 32 teams.
這個籃球聯盟是由三十二支隊伍所組成的。

5. compliance [kəm`plaɪəns] *n.* 遵從；順從

Do the parts show compliance with all of our regulations?
這些零件有遵從我們所有的規定嗎？

6. randomly [`rændəmlɪ] *adv.* 隨機地；任意地

We randomly chose five products to inspect.
我們隨機抽樣五件產品來檢驗。

7. economical [ˌikə`nɑmɪkl̩] *adj.* 經濟的；節省的

Sally is trying to be very economical until she gets her paycheck.
莎莉在拿到薪水前試著非常節省。

8. reputable [`rɛpjətəbl̩] *adj.* 聲譽好的；可尊敬的

If you want to buy a used motorcycle, you should go to a reputable shop.
如果你想要買一台二手摩托車，你應該去有信譽的商店。

Biz Focus

★ final quality control 成品檢驗（常縮寫成 FQC）

★ in favor of 付予

Language Corner

Trying to Save Money

I hope working with you will make it as economical as possible for us.

> *You Might also Say . . .*
>
> ▪ I hope hiring you as our agent can help us conserve as much money as possible.

Unit Four
Information for Quoting
報價所需資訊

報價太高搶不到生意，報價太低會虧本。能否確實報價和客戶所提供的資訊是否完整有關，如不完整應立即向客戶反應、澄清疑點，不可預設立場妄下決定。

✅ Learning Goals

1. 向客戶澄清疑點時，需充分展現專業上的自信，勿讓對方覺得是外行人在提問

2. 有時客戶會指定使用某廠牌的零組件，如欲採用代替品，需先取得對方同意後再行報價

3. 報價前需先請客戶提供品管檢驗標準及程序

Before We Start 實戰商務寫作句型

Topic: Finding Out the Latest 詢問最新消息

- (In order) to proceed / go ahead / advance with the obtaining / acquiring of quotes for . . ., we'll need the following [information / data / details / specifics . . .] . . .

  為進一步取得……的報價，我們將需要下列 [資料 / 細節……]……

- When will the [product specifications / test procedure / shipping schedule . . .] be available / accessible?

  何時才能拿到 [產品規格 / 檢驗程序 / 運貨時間表……]？

- Can you enlighten me on / notify (inform / apprise) me of what [test equipment / packaging / dimensions . . .] will be required / requested / needed?

  可以請你告訴我會需要哪種 [檢驗設備 / 包裝 / 尺寸……] 嗎？

- If you have a milestone chart / schedule / timetable of some sort (kind) [written down / drawn up . . .] for this project, we would appreciate a copy / duplicate / facsimile.

  針對這個案子，如果你有某種 [書面 / 草擬的……] 里程碑表 / 時間表，請給我們一份副本。

- Should you have any more data that might be of interest / that's pertinent (relevant), by all means / be sure to / definitely forward it to me.

  如果你還有更多相關的資料，請務必轉寄給我。

- Before [formalizing / finalizing / approving . . .] the quote, we need to confirm / certify / verify . . .

  在 [敲定 / 同意……] 這份報價以前，我們需要確認 / 證實……

A. Request for Information 索取資訊

#1 Richard 在確定可以供應客戶所需的產品後,便向客戶要詳細資料以準備報價。

To: Luke Lerer
From: Richard Barlow
Date: December 24
Subject: Info Needed for Quoting

Dear Mr. Lerer,

We here at FETS **are appreciative of**[1] the opportunity to serve you. I am confident that we can provide **stellar**[2] service to you in both having tooling made and finding a manufacturer for your products.

To proceed with the obtaining of quotes for the tooling and manufacture of the parts (handles), we'll need the following information:

1. Drawings of the plastic parts
2. Samples of the parts (if available)
3. Specifications of the parts (material, dimensions, colors, etc.)
4. A forecast of the quantity requirements for the first year
5. Shipping schedules for the first year for each part
6. Packaging specifications for the finished products

We'll anticipate receiving the above info **before long**[3] so that we can **initiate**[4] the process.

Thank you for your holiday wishes, and the same to you.

Sincerely,
Richard Barlow

中譯

#1
收件人:路克・雷勒
寄件人:理查・巴洛
日期:十二月二十四號
主旨:報價所需資料

親愛的雷勒先生:

FETS 感謝能有這個機會為您服務。我有信心我們在製造模具及尋找貴公司產品的製造商方面均能提供您一流的服務。

為進一步取得模具和零件(把手)製造的報價,我們將需要下列資料:

1. 塑膠零件的圖面
2. 零件的樣品(如可提供的話)
3. 零件的規格(材料、尺寸、顏色等)
4. 預計第一年的需求量
5. 第一年各零件的出貨時間表
6. 成品的包裝規格

我們期待不久之後能收到以上的資料這樣我們才能開始報價的程序。

謝謝您的佳節祝福,也祝您佳節愉快。

祝 商安
理查・巴洛

Vocabulary and Phrases

1. **be appreciative of** 感謝
   Steve was very appreciative of his girlfriend's offer to cook him dinner.
   史提夫非常感謝他的女友願意為他做晚餐。

2. **stellar** [ˈstɛlɚ] *adj.* 一流的;主要的
   Jennifer has been a stellar salesperson since her first day on the job.
   珍妮佛從她上班的第一天開始就一直是一位頂尖的銷售員。

3. **before long** 不久之後
   If you keep showing up to work late, you'll be fired before long.
   如果你繼續上班遲到,不久之後你就會被開除。

4. **initiate** [ɪˈnɪʃiˌet] *v.* 開始
   The company initiated talks about hiring a design firm.
   該公司開始談到要雇用一間設計公司。

#2 產品有用到印刷電路板，組裝完成後需全檢，而這會增加人工成本及儀器費用，故 Richard 特別跟客戶詢問測試方法及儀器等問題。

To: Luke Lerer
From: Richard Barlow
Date: 1/5
Subject: Board Questions

Dear Mr. Lerer,

　　We did receive the plastic handle drawings this week, and thank you for the additional materials you sent today. We will have a quote for the entire assembly to you next week.

　　When will board artwork and specifications be available? Can you **enlighten** me **on**[5] what test procedure and equipment will be required?

　　If you have a milestone chart **of some sort**[6] written down for this project, we would appreciate a copy. Should you have any more data that might be of interest, **by all means**[7] forward it to me.

Thanks and best regards,
eom

中譯

#2

收件人：路克・雷勒
寄件人：理查・巴洛
日期：一月五號
主旨：電路板的問題

親愛的雷勒先生：

　　我們確實已於本週收到塑膠把手的圖面，並感謝您今天另外寄來的資料。我們下週會提供您整組的報價。

　　何時才能拿到電路板的線路設計圖和規格呢？可以請您告訴我會需要什麼樣的檢驗程序和設備嗎？

　　針對這個案子，如果您有某種書面的里程碑表，請給我們一份副本。如果您還有更多相關的資料，請務必轉寄給我。

感謝 謹上
完畢

---

enlighten sb on sth 向某人解釋某事；使某人明白某事
Could you enlighten me on the company's new policy?
你可以和我解釋一下公司的新政策嗎？

of some sort 某種
We need to see identification of some sort to allow you to use a credit card.
我們需要查看某種證件才能讓你使用信用卡。

by all means 務必
If you have any questions, by all means give me a call.
如果你有任何問題，務必要打電話給我。

Language Corner

Asking for Important Data

We'll anticipate receiving the above info before long so that we can initiate the process.

You Might also Say . . .

- We need to get the above information shortly so that we can get things moving.
- Please send us the above data in a timely manner so we can get down to business.

B. Product Requirements 產品要求

Richard 報價時，發現市面上找不到客戶指定要用的組件，於是寫信問客戶是否可用代替品及其他週邊設備的問題。

To: Andy Marvel
From: Richard Barlow
Date: 10/17
Subject: Quote Update

Dear Mr. Marvel,

Thank you for giving us the opportunity to quote for you. We are working on your requests, and we have the following questions and comments:

A. Printek microprinters are not available for export from Taiwan because they are not manufactured in Taiwan. Taiwanese companies that need microprinters import them from Japan.

Fortunately, we have discovered a supplier of microprinters that is **based**[1] in Hong Kong and manufactures in China. Their units are built to Printek specifications, and the supplier has the capacity to meet your requirements.

B. Before **formalizing**[2] the quote, we also need to confirm:

1. Will you require an interface* with the printer?

2. There is a panel-mount model* available, which I think would be **suitable**[3] for what I imagine your **application**[4] to be. Are you interested?

C. We received your sample LED* yesterday, and we are in the process of collecting quotes. There are many LED manufacturers in Asia. Some would consider getting the LEDs from China because of the **relatively**[5] lower costs there, but Taiwan has a pretty good **reputation**[6] for producing quality LEDs. However, we would be able to present you with both quotes next week for your consideration.

Sincerely,
Richard Barlow

中譯 收件人：安迪‧馬維爾

寄件人：理查‧巴洛

日期：十月十七號

主旨：報價最新進展

親愛的馬維爾先生：

謝謝您給我們這個機會為您報價。我們正在處理您的需求，下列是我們的一些問題和意見：

A. Printek 微型印表機無法自台灣出口，因為它們不是在台灣製造的。需要微型印表機的台灣公司都是從日本進口。

所幸我們找到一間總公司在香港而在大陸生產的微型印表機供應商。他們的產品是依照 Printek 的規格製造的，而該供應商也有能力符合您的需求。

B. 在敲定這份報價之前，我們還需要確認：

1. 您需要印表機的介面嗎？

2. 有一種裝有面板的機型，我認為很適合在我想像中你們產品的應用。您有興趣嗎？

C. 我們昨天收到你們發光二極體（LED）的樣品，而我們正在收集報價。亞洲有很多發光二極體的製造商。有些公司會考慮從大陸採購發光二極體，因為那邊的成本相對較低，但台灣在生產高品質的發光二極體方面享有盛名。無論如何，下星期我們會提供兩種報價讓您斟酌。

祝 商安

理查‧巴洛

Vocabulary and Phrases

1. **base** [bes] *v.* 以……為基地、總部

   The company does business worldwide but is based in New York.
   該公司的生意遍布全球，但以紐約為其總部。

2. **formalize** [`fɔrml̩ˌaɪz] *v.* 使正式；使形式化

   We've got to formalize your hiring by having you sign this contract.
   我們必需讓你簽這份合約以正式聘雇你。

3. **suitable** [`sutəbl̩] *adj.* 適當的；合適的

   Jeff's personality makes him suitable for the sales position.
   傑夫的個性適合當業務員。

4. **application** [ˌæpləˋkeʃən] *n.* 應用；適用

   The Swiss Army knife has many different applications.
   瑞士刀有很多不同的用法。

5. **relatively** [`rɛlətɪvlɪ] *adv.* 相對地；比較而言

   Rent in the country is relatively cheap when compared to the cost of city housing.
   和在城市居住的花費相比，鄉下的房租相對便宜。

6. **reputation** [ˌrɛpjəˋteʃən] *n.* 信譽；聲望

   This product has a reputation for having good quality.
   這項產品以高品質著名。

Biz Focus

★ **interface** 連接器；界面

★ **panel-mount model** 裝有面板的機型

★ **LED** 發光二極體（為 Light Emitting Diode 的簡寫）

Language Corner

Giving a Reputation

Taiwan has a pretty good reputation for producing quality LEDs.

You Might also Say . . .

- Taiwan's production of quality LEDs is held in high esteem.
- Taiwan is renowned for its production of high-grade LEDs.
- Taiwan has gained a reputation as the producer of well-made LEDs.

Unit Five
Quote Formalization
正式報價

可以針對產品或服務提供報價。報價要儘可能詳細，讓客戶清楚了解價格、交期、付款條件等。而為避免原物料的價格波動，報價亦需註明有效期限。

✔ Learning Goals

1. 模具的報價除了模具費外，還需註明模穴數、開模所需時間及使用壽命等

2. 樣本數大小、是否需使用特殊的儀器等，都會影響到品管檢驗的報價

3. 報錯價時需在第一時間向客戶道歉並更正

Before We Start 實戰商務寫作句型

Topic 1: Discussing Quotation Terms　討論報價條件

- The [shipping / inspection / consultation . . .] cost / price / charge / fee is open to discussion / negotiable / flexible.
  [運送 / 檢驗 / 諮詢……] 的費用是可以協商 / 有彈性的。

- This quotation is valid / good / in force for (a period of time).
  這份報價的有效期為（一段時間）。

- This quote is [based on the assumption / grounded on the fact / founded on the knowledge . . .] that . . .
  這份報價是依據……的 [假設 / 事實 / 認知……] 而來。

- I'm looking forward to receiving / awaiting the receipt of / anxious to receive your acceptance / approval of this offer.
  我期望你能接受 / 同意這份報價。

Topic 2: Apologizing for Inaccurate Quotes　為報錯價道歉

- I reviewed / looked over / reexamined this quote today, and I discovered why our price looked so good / enticing / appealing.
  我今天重新檢視了這份報價，我發現到為什麼我們的價錢看起來這麼吸引人了。

- I'm afraid / I regret to inform you / I hate to tell you that we made a(n) oversight / blunder / mistake / error.
  很抱歉（要通知你）我們犯了一個錯誤。

- I'm very sorry / Please accept my apology / Please excuse (forgive) me for the [inaccuracy / mix-up / carelessness . . .].
  我要為此 [錯誤 / 混亂 / 疏忽……] 向你道歉 / 請求你的原諒。

- Please don't think this is a routine / normal / common / frequent occurrence.
  請勿認為這種情形時常發生。

IBK 欲透過 FETS 在台採購鋅壓鑄品，於是 Richard 寫信向其報價，並溝通一些有關開模及製程的問題。

○ ○ ○

To: William Kay
From: Richard Barlow
Date: 3/5
Subject: Three Project Quotes
Attachment: kay_quotation.doc

Dear Mr. Kay,

Quotations for projects A-00518, A-00497, and A-00498 are attached. Please note that on item A-00518, our vendors would not quote an **adjustable**[1] tool as you requested for **a myriad of**[2] reasons, from poor part quality to short tool life to technical difficulties in the tool **construction**.[3] Please let us know if you still would like a quotation for low-pressure casting.*

Best regards,
eom

Attachment

Updated March 5

FAR EAST TECHNICAL SERVICES

We are pleased to offer you the following quotations for zinc die-cast parts, projects A-00518, A-00497, and A-00498:

Project	Drawing #	Tooling Cost	Cavity Number	Part Cost
A-00518	IBK-160 / 82	6,150	1	4.39
A-00497	Part 1	7,350	1	4.49
A-00498	N / A	6,150	1	5.31

Costs are shown in USD.

Delivery[4] time: 80 days.

Terms: L / C at sight* or 30% **down payment**;[5] 40% at first shots;* 30% when tooling is approved.

Tool life: The tooling replacement cost is **negotiable**.[6] Under normal circumstances, there is no additional charge for replacement tooling.

This quotation is valid for 60 days.

中譯

收件人：威廉‧凱

寄件人：理查‧巴洛

日期：三月五號

主旨：三個案子的報價

附件：kay_quotation.doc

親愛的凱先生：

　　附件為 A-00518、A-00497 和 A-00498 案子的報價。請注意在 A-00518 這項中，我們的廠商不願依您所要求的針對可調整式模具來報價，原因有很多像是零件品質不佳、模具壽命太短、開模技術上的困難等。如果您仍需要低壓鑄造的報價請讓我們知道。

謹上
完畢

📎 附件

三月五號更新

FAR EAST TECHNICAL SERVICES

我們很高興能針對鋅壓鑄零件的案子 A-00518、A-00497 和 A-00498，提供您以下的報價：

案子	圖面編號	模具費	模穴數	部品單價
A-00518	IBK-160 / 82	6,150	1	4.39
A-00497	Part 1	7,350	1	4.49
A-00498	無	6,150	1	5.31

＊ 以美金計價。

交貨期：八十天。

付款條件：即期信用狀或頭期款付百分之三十；第一次試模付百分之四十；模具驗收後付百分之三十。

模具壽命：替換模具的費用是可以協商的。在正常的情況下，替換的模具不需額外支付費用。

這份報價的有效期為六十天。

Vocabulary and Phrases

1. **adjustable** [ə`dʒʌstəbl̩] *adj.* 可調整的
   He took out his adjustable wrench and worked on his motorcycle.
   他拿出了活動扳手來修理他的摩托車。

2. **a myriad of** 無數的；大量的
   There's a myriad of restaurants in the downtown area.
   市區有很多的餐廳。

3. **construction** [kən`strʌkʃən] *n.* 建造
   The construction of the house took eight weeks.
   建造那間房子花了八個星期。

4. **delivery** [dɪ`lɪvərɪ] *n.* 交貨；遞送
   The delivery of the pizza will take 45 minutes.
   外送比薩需花四十五分鐘。

5. **down payment** 分期付款的頭期款
   One month after the wedding, the couple made a down payment on a new house.
   婚禮過後一個月，那對夫妻付了新房子的頭期款。

6. **negotiable** [nɪ`goʃɪəbl̩] *adj.* 可協商的
   The shipping cost for large orders is negotiable.
   大量訂單的運費是可以協商的。

Biz Focus

★ **low-pressure casting** 低壓鑄造（品）

★ **L / C at sight** 即期信用狀

★ **first shot** 第一次試模

Language Corner

Describing What Is Usual

Under normal circumstances, there is no additional charge for replacement tooling.

You Might also Say . . .

- In a normal situation, there is no surcharge on replacement tooling.
- Usually / Typically / Ordinarily, you don't need to pay extra to replace the tooling.

B. Service Quote 服務報價

#1 TEK 欲在台採購車床加工的機械零件，於是請 FETS 針對代為作出貨檢驗一事來提供報價。

To: Kevin Green
From: Richard Barlow
Date: 2/27
Subject: Re: Lot Inspections

Dear Kevin,

Regarding the lot inspections of machined metal parts as we discussed in yesterday's meeting, we can inspect a shipment of parts for US$265.

This quote is good for all parts similar in **complexity**[1] to the samples of the **shields**[2] you gave us. It is based on the **assumption**[3] that the inspections can be suitably **executed**[4] with high-quality handheld measuring devices and that the sample size is 125 pieces. Also, we understand that you will supply go / no-go gauges* for all holes and thread gauges* for all threads, and that no environmental conditioning or material analysis is required as part of the inspections.

I'm looking forward to receiving your acceptance of this offer.

Sincerely,
Richard

中譯

#1

收件人：凱文・葛林
寄件人：理查・巴洛
日期：二月二十七號
主旨：回覆：批次檢驗

親愛的凱文：

關於我們昨天在會議中討論的金屬機械零件批次檢驗，我們檢驗一批零件出貨的價格是兩百六十五元美金。

這份報價適用於具有和你給我們的防護罩樣品相似複雜度的所有零件，並假定可使用精密手握式測量工具適當地檢驗，且樣品數為一百二十五件。再者，我們知道所有的孔洞及螺紋你們都會提供通過／不通過量規和牙規，同時檢查也不需包含特殊環境檢驗或物料分析。

我期望你能接受這份報價。

祝 商安

理查

Vocabulary and Phrases

1. complexity [kəmˈplɛksətɪ] *n.* 複雜（性）

The complexity of the question caused the man to be confused.
這個問題的複雜性使那名男子感到困惑。

2. shield [ʃild] *n.* 防護罩

They inspected the shields and discovered they weren't thick enough.
他們檢查了防護罩，發現它們不夠厚。

3. assumption [əˈsʌmpʃən] *n.* 假定；認為（assume *v.*）

Valerie made the assumption that Will would be on time to meet her at the movie theater.
瓦萊麗假定威爾能準時與她在電影院碰面。

4. execute [ˈɛksɪˌkjut] *v.* 執行；實行

The inspection plan was executed very well.
那個檢驗計畫執行得很成功。

5. in anticipation of 期待；預期

Sam has been working hard in anticipation of a promotion.
山姆努力工作期望能升職。

#2 FETS 為爭取與 CPS 公司的長期合作、加上出貨的工廠不遠，於是願意免費替客戶作第一次的出貨檢驗。

To: Stephen Cameron
From: Richard Barlow
Date: 1/10
Subject: Golf Clubs

Dear Mr. Cameron,

Thank you for giving us the opportunity to quote the job as detailed in your letter dated 1/6.

Because we'd like to take this chance to expand our product expertise to include sporting goods (and because the factory is located nearby), we'll inspect this shipment of golf clubs for you at no charge **in anticipation of**[5] building a long-term relationship.

If this arrangement is acceptable to you, please forward the artwork and color chips* along with any other samples pertinent to the inspections. I assume that you will also be sending us specifications, an inspection plan, and particulars regarding the factory.

Also, please confirm that the manufacturer will handle all shipping matters.

We're hoping to start a mutually beneficial business relationship with you.

Your sincerely,
Richard Barlow

中譯

#2

收件人：史帝文‧卡麥隆
寄件人：理查‧巴洛
日期：一月十號
主旨：高爾夫球桿

親愛的卡麥隆先生：

感謝您給我們機會針對您在一月六號來信中詳述的工作向您報價。

由於我們想藉由這次機會將我們的產品經驗擴展至運動用品（也因為工廠就在附近），所以我們將免費替你們檢驗這批高爾夫球桿，以期能與你們建立長期的合作關係。

如果這樣的安排你們可以接受，請轉寄產品包裝設計圖、色卡和任何其他與檢驗相關的樣品給我們。我假定您也會寄給我們規格、檢驗計畫和有關工廠的細節。

另外，請確認製造商會處理所有的出貨事宜。

我們希望能與你們開始一段互利的商業關係。

謹上
理查‧巴洛

▶ Biz Focus

★ **go / no-go gauge** 通過 / 不通過量規
　檢驗大量工件的內、外徑時，可以快速判別是否在允許的公差範圍內。只要 go 端通得過而 no go 端通不過，該工件就屬合格。

★ **thread gauge** 牙規（測量螺紋是否在公差範圍內）

★ **color chip** （標準）色紙、色卡

Language Corner

Establishing a Business Relationship

We're hoping to start a mutually beneficial business relationship with you.

You Might also Say . . .

- We have high hopes of starting a mutually profitable partnership with you.

- We would like to establish a business relationship that benefits both of us.

C. Revised Quote 修正報價

Richard 在發現報錯價後立即向客戶道歉並補上正確的報價。

To: Kevin Green
From: Richard Barlow
Date: 1/27
Subject: Quote **Inaccuracy**[1]

Dear Kevin,

I reviewed this quote today, and I discovered why our price looked so **enticing**.[2] I'm afraid that we made an **oversight**.[3] I'll save you the trouble of listening to excuses and just sincerely offer my apology. I'm very sorry for the mistake. Please don't think this is a routine **occurrence**.[4]

We should have quoted the following:

20.2 x 5 cm Mylar* circuit

Qty	Unit Price (USD)
25K	0.88
50K	0.84
100K	0.80
200K	0.78

As you know, flex circuits* are an area in which we **know the ropes**.[5] We are, of course, very **intrigued**[6] by doing business with you. I hope that you find this quote to be acceptable. Let me mention again that as the volume of work **builds up**,[7] I will readily expand our operation as necessary to fully support your business. And as the quantity increases, I'm confident that the unit cost can be reduced.

Again, I apologize for the blunder on our part.

Kind regards,
Richard Barlow

中譯

收件人：凱文・葛林
寄件人：理查・巴洛
日期：一月二十七號
主旨：報價錯誤

親愛的凱文：

我今天重新檢視了這份報價，我發現到為什麼我們的價錢看起來這麼吸引人了。很抱歉我們有所疏失。沒有任何藉口，我只想向您真摯地表達我的歉意。我很抱歉犯了這個錯誤。請勿認為這種情形時常發生。

我們的報價應如下列：

20.2 x 5 公分 美拉電路

數量	單價（美金）
兩萬五千件	0.88
五萬件	0.84
十萬件	0.80
二十萬件	0.78

如您所知，柔性電路板是我們很在行的一個領域。我們當然非常有興趣與您合作。我希望您會覺得這份報價是可以接受的。容我再提一次，當工作量增加時，如果有需要我會立即擴充我們的營運以全力支援您的生意。而且當數量增加時，我確信單價也會跟著降低。

我要再次為我們這邊所犯的錯向您道歉。

謹上
理查・巴洛

Vocabulary and Phrases

1. inaccuracy [ɪnˋækjərəsɪ] *n.*
錯誤;不正確(inaccurate *adj.*)

The reader noticed an inaccuracy in the newspaper column.
那名讀者注意到這篇報紙專欄有一個錯誤。

2. enticing [ɪnˋtaɪsɪŋ] *adj.* 引誘的;迷人的

The junk food was enticing, but he couldn't eat it because he was on a diet.
垃圾食物很吸引人,但是他在節食不能吃。

3. oversight [ˋovɚ͵saɪt] *n.* 失察;疏忽出錯

The oversight in the report cost the company one million dollars.
那份報告的疏失使公司損失了一百萬元。

4. occurrence [əˋkɝəns] *n.* 事件;發生

Getting to meet the president is a very rare occurrence.
和總裁見面是很少見的事。

5. know the ropes 懂門道;在行

When it comes to selling computer products, Danielle really knows the ropes.
當提到賣電腦產品時,丹妮爾真的很在行。

6. intrigued [ɪnˋtrigd] *adj.* 好奇的;被迷住的

I want to invest in the stock market, and I'm intrigued by this technology company.
我想要投資股市,而我對這家科技公司非常感興趣。

7. build up 逐漸變大、多

The crowd noise always builds up near the end of the basketball game.
接近籃球比賽終場時群眾的喧鬧聲總會變大。

Biz Focus

★ Mylar 美拉(一種聚酯薄膜,為商標名)

★ flex circuit 柔性電路

Language Corner

1. Reassuring a Customer

Please don't think this is a routine occurrence.

You Might also Say . . .

- Please don't assume that this happens often.
- Please don't think events like this occur all the time.

2. Offering a Quantity Discount

As the quantity increases, I'm confident that the unit cost can be reduced.

You Might also Say . . .

- As the volume becomes greater, I'm certain that the unit cost will come down.
- If you order larger quantities, I'm sure that the price of each unit will be decreased.

Unit Six
Quote Follow-up
報價後續追蹤

報價一段時間後，如還沒收到客戶的回音，可發一封追蹤信函，確認對方是否有收到報價、詢問其對報價的看法，並表達「如不滿意仍可進一步協商」的立場，以期能促成該筆交易。

✅ Learning Goals

1. 為爭取新客戶，要儘量壓低第一次報價，以建立長期合作關係為目標

2. 客戶不滿意報價時，需設法得知對方的理想價格，以便適當調整、增加成交的機率

Before We Start 實戰商務寫作句型

Topic: Pursuing a Partnership 尋求合作關係

* As I'm sure / I trust (think) you'll remember, on (date), we offered
  you / presented (supplied / provided) you with a quotation for
  (product).
  我想您會記得在（日期），我們有提供您（某產品）的報價。

* We haven't heard anything back / received any feedback (reply /
  response) from you since [then / that time / we last spoke . . .].
  從 [那時起 / 我們上次談話後……]，我們一直沒有聽（收）到您的回音。

* As I recall that you said the quote warranted / required / called
  for further [discussion / negotiation / adjustment . . .], I'm curious
  about your current [level of interest in / thoughts on / opinion of . . .]
  the plan.
  由於我記得您說過這份報價需要進一步的 [討論 / 協商 / 調整……]，我想知道您
  目前對這個案子的 [意願 / 想法……]。

* Please update us / send us an update / bring us up to date at
  your earliest convenience / whenever it is convenient / when
  you get the chance.
  您方便的時候，請（儘早）讓我們知道最新狀況。

* We agreed to lower / reduce / decrease the [shipping charge /
  consultation fee . . .] in (with) the hope of / with an eye towards
  establishing / building / founding a long-term business relationship.
  我們同意減少 [運費 / 諮商費……]，以期建立長遠的商業關係。

* We will forgo / pass up / give up our [commission / profit . . .] so
  we can supply you with (product) for your [requested / desired /
  preferred . . .] price.
  我們將放棄我們的 [佣金 / 利潤……]，依您所 [要求 / 想要 / 傾向……] 的價格
  提供您（某產品）。

A. Keeping in Touch 保持聯繫

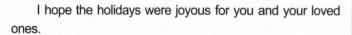

To: Ben Milton
From: Vincent Johnson
Date: January 6
Subject: **Insurance**[1] Quote Follow-up

Dear Mr. Milton,

I hope the holidays were joyous for you and your loved ones.

As I'm sure you'll remember, on November 20, we offered you a quotation for life insurance. We haven't heard anything back from you since then, and because I **recall**[2] that you said the quote **warranted**[3] further discussion, I'm curious about your current level of interest in the plan.

I know you're aware that you take an unnecessary risk each day that you **put off**[4] purchasing life insurance. Please update us at your earliest convenience. We want to get to work on finalizing a plan to offer protection for you and your family.

Sincerely,
Vincent Johnson
Sales Representative
Safety Net Life Insurance

中譯

收件人：班‧彌爾頓
寄件人：文森‧強森
日期：一月六號
主旨：保險報價的後續追蹤

親愛的彌爾頓先生：

我希望您和您的摯愛都有個愉快的假期。

我想您會記得在十一月二十號，我們有提供您壽險的報價。從那時起我們一直都沒有聽到您的回音。因為我記得您說過這份報價需要進一步的討論，所以我想知道您目前對這個案子的意願。

我明白您也意識到延遲購買壽險，您每天都將冒不必要的風險。您方便的時候，請儘早讓我們知道最新狀況，我們想著手完成一個能提供您和您家人保障的計畫。

祝 商安
文森‧強森
業務代表
Safety Net 壽險公司

Vocabulary and Phrases

1. insurance [ɪnˈʃʊrəns] *n.* 保險；保險業；保險金額
You're not allowed to drive a car without insurance.
沒有保險你就不准開車。

2. recall [rɪˈkɔl] *v.* 記得；回想
Jane couldn't recall where she had put the documents.
珍不記得她把文件放在哪裡了。

3. warrant [ˈwɔrənt] *v.* 使有（正當）理由；成為……的根據
The manager thought Alison's performance in the interview warranted a second meeting.
經理認為根據艾莉森在面試時的表現，她可獲得第二次的會面。

4. put off 延遲；拖延
I put off paying my phone bill until I got my paycheck.
我一直拖到領薪水才付我的電話帳單。

5. in the hope of 懷著……的希望
Paul moved to the country in the hope of living a quiet life.
保羅搬到鄉下希望能過著寧靜的生活。

6. forgo [fɔrˈgo] *v.* 放棄；棄絕
She forwent her planned vacation in order to get the project completed on time.
為了讓這個案子準時完成，她放棄已計畫好的假期。

B. Renegotiation 再次協商

To: James Franklin
From: Sam Cunningham
Date: 3/25
Subject: Shipping Charge

Dear Mr. Franklin,

Thank you for your fax, which we received today.

We reviewed the shipping charge with our vendor, and he agreed to lower it **in the hope of**[5] establishing a long-term business relationship. We will also **forgo**[6] part of our commission towards this goal so we can supply you with the **apparel**[7] for your requested price of US$2,100 per unit. Please appreciate that we anticipate simply making enough money to cover our expenses on this deal. Hopefully, it will serve as a sign of good faith and help to get us on our way towards a **prolonged**[8] partnership.

Please let me know if I can do anything else to help in this matter.

Best regards,
Sam Cunningham
Sales Manager
Featherweight Clothing

收件人：詹姆士·法蘭克林
寄件人：山姆·康尼翰
日期：三月二十五號
主旨：貨運費用

親愛的法蘭克林先生：

謝謝您今天傳來的傳真，我們收到了。

我們和供應商重新檢討了運費，而他也同意減少費用，以期建立長遠的商業關係。為了這個目標，我們也將放棄部分的佣金，依您所要求的單價兩千一百元美金，提供您該批服裝。請體諒我們只期望能賺取足夠的錢來支付這筆交易的支出，但願這樣可以表示我們的誠意，讓我們朝建立長遠夥伴關係的目標邁進。

如果有其他我能幫得上忙的地方，請讓我知道。

謹上
山姆·康尼翰
業務經理
Featherweight 服裝公司

---

7. **apparel** [ə`pærəl] *n.* 衣服；服飾
She went to the sporting goods store to buy some workout apparel.
她去運動用品店買了一些運動服飾。

8. **prolonged** [prə`lɔŋd] *adj.* 延長的；特別長的
He decided to take a prolonged vacation on the island.
他決定在這個島上度長假。

Language Corner

Offering Further Help

Please let me know if I can do anything else to help in this matter.

You Might also Say . . .

- Please let me know if I can do more to support this.
- Feel free to tell me if there is anything I can do to be of assistance.
- Don't hesitate to ask for more help.

Chapter Five

Production /
Manufacturing

Unit One
Product Evaluation
產品評估

開發新產品前需先針對市面上同性質的產品進行價格、功能、競爭力等的調查。在確定自家產品具有競爭力,且有明確的市場定位後,再投入資金和人力進行後續工作。

 Learning Goals

1. 分析競爭對手的產品,並客觀評估自身的優缺點

2. 參加國際商展有助了解市場對自家產品的接受度

3. 新產品要設法取得專利,以保有市場競爭力

Before We Start 實戰商務寫作句型

Topic 1: Reviewing a Product 檢視產品

◆ The product appears to have several [performance / design / usability . . .] shortcomings / flaws / weaknesses / faults.
那項產品似乎有一些 [性能 / 設計 / 使用性……] 方面的缺陷。

◆ Could you please have somebody forward a working sample to me for a [technical evaluation / comparative analysis / product survey . . .]?
你可不可以請人轉寄一份供試樣品給我，讓我們作 [技術評估 / 比較分析 / 產品調查……]？

Topic 2: Introducing Your Product 介紹自家產品

◆ (Product) is a(n) revolutionary / innovative / pioneering / groundbreaking / never-before-seen [device / system / tool . . .].
（某產品）是一項創新的 / 首創的 [裝置 / 系統 / 工具……]。

◆ (Product) improves / enhances / upgrades [hygiene / efficiency / sound quality / comfort . . .] by . . .
（某產品）藉由……來改善 / 提升 [衛生 / 效率 / 音質 / 舒適度……]。

◆ (Product) can be [installed / set up / put together / assembled . . .] in seconds / moments / minutes / no time.
（某產品）可快速 [安裝 / 組裝……]。

◆ (Product) reduces / lowers / lessens [water / energy / electricity / fuel / gasoline . . .] consumption by [up to / more than / at least . . .] (percentage).
（某產品）能減少 [多達 / 超過 / 至少……]（百分之幾）的 [水力 / 能源 / 電力 / 燃料 / 汽油……] 消耗。

Maples 公司新研發一款自動控水裝置，在產品上市前先與 OEM 工廠交換有關競爭產品的資訊。

TO: Gary Yang
FROM: Todd Andrews
DATE: March 30
SUBJECT: **Automatic**[2] Water-control Device

Dear Gary,

Thank you for your e-mail today regarding the International water-control device. Although the product appears to have several performance **shortcomings**,[3] could you please have somebody forward a working sample* to me for a technical evaluation? We **urgently**[4] need it, as our engineering department is finalizing the design work and the tooling orders will be placed in two weeks. Thanks.

Additionally, please find the attached overview of our HandsOff™ product (Attachment #1) and a **comparison**[5] between HandsOff™ and an **add-on**[6] device by Handy (Attachment #2). The company clearly uses our design and is also attempting to market this unit to the dental **trade**[7] here in the United States. We are just about to proceed with legal action.

I appreciate your updates on our competitor's product. Tomorrow, we will have a meeting reviewing HandsOff™. I will let you know what comes out of it.

Kind regards,
Todd Andrews
Vice President
New Business Development

中譯 收件人：楊蓋瑞

寄件人：陶德・安德魯斯

日期：三月三十號

主旨：自動控水裝置

親愛的蓋瑞：

　　謝謝你今天有關寰宇牌控水裝置的來信。雖然該產品似乎有一些性能方面的缺陷，你可不可以還是請人轉寄一份供試樣品給我，讓我們作技術評估？我們急著要，因為我們的技術部門正要對設計做最後的確認，而且兩個星期後模具也要下單了。謝謝。

　　此外，請參見我們 HandsOff™ 產品的簡介（附件一）以及 HandsOff™ 與 Handy 公司生產的附加裝置比較（附件二）。該公司很明顯地是使用我們的設計，並試圖將產品賣給在美國這裡的牙醫業。我們正要採取法律行動。

我很感激你提供有關我們競爭對手產品的最新資訊。明天我們會開會檢討 HandsOff™，我將會讓你知道會議的結果。

謹上
陶德・安德魯斯
副總裁
新業務發展部門

Vocabulary and Phrases

1. **competitor** [kəm`pɛtətɚ] *n.* 競爭者；對手
   We've been competitors for ages.
   我們長期以來都是競爭對手的關係。

2. **automatic** [ˌɔtə`mætɪk] *adj.* 自動的；自動裝置的
   Jill bought a camera with an automatic focus.
   吉兒買了一台有自動對焦的相機。

3. **shortcoming** [`ʃɔrtˌkʌmɪŋ] *n.* 缺點；短處
   My job's only shortcoming is that it's too far from my house.
   我的工作的唯一缺點就是離我家太遠。

4. **urgently** [`ɝdʒəntlɪ] *adv.* 緊急地；急迫地
   Philip's boss said he urgently wanted to meet him.
   菲力普的老闆說要馬上見他。

5. **comparison** [kəm`pærəsn̩] *n.* 比較；對照
   Molly did a comparison of the two items before deciding to buy the cheaper one.
   莫莉比較了一下這兩項產品才決定買其中比較便宜的。

6. **add-on** [`æˌdɑn] *adj.* 附加的
   There is a wide variety of add-on components for your computer.
   你的電腦可以加裝各式各樣的組件。

7. **trade** [tred] *n.* 行業
   He's been in the insurance trade for more than 30 years.
   他在保險業已經超過三十年了。

Biz Focus

★ working sample （功能完整、可供測試及作技術評估的）供試樣品

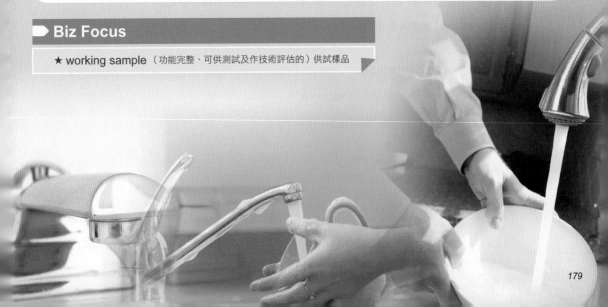

Attachment #1

HANDSOFF™
AUTOMATIC WATER-CONTROL DEVICE

Model HO-1A

HANDSOFF™ HELPS MEET TODAY'S **STRINGENT**[1] HEALTH-CODE REQUIREMENTS

- **Revolutionary**[2] add-on device turns an ordinary faucet into an electronic one.

- **Patented**[3] ultrasonic sensing technology with a 32-bit **microprocessor**[4] automatically **activates**[5] and deactivates water flow.

- Improves **hygiene**[6] by eliminating contact with faucet handles.

- Installs in seconds — no tools, no **plumbers**,[7] no electricians.

- Reduces water and energy **consumption**[8] by up to 60%.

- Powered by a standard alkaline 12-volt battery (nine-to-12-month battery life).

- Two modes of operation — manual and automatic.

- Includes three standard **adapters**.[9]

- **Theft-proof**[10] model available (TF-1B).

  附件一

HANDSOFF™
自動控水裝置
HO-1A 型號
HANDSOFF™ 有助達到今日嚴格的健康規範要求

- ✅ 創新的附加裝置讓一般的水龍頭變成電動水龍頭。

- ✅ 取得專利的超音波感應技術與三十二位元的微處理器能自動開啟、關閉水流。

- ✅ 藉由減少接觸水龍頭把手來改善衛生。

- ✅ 快速安裝 — 不需要工具、水管工人或電工。

- ✅ 減少多達百分之六十的水力與能源消耗。

- ✅ 用一般鹼性十二伏特的電池來發電（電池壽命為九到十二個月）。

- ✅ 兩種使用模式 — 手動與自動。

- ✅ 包含三個標準的轉接器。

- ✅ 亦有防盜型的機種（TF-1B）。

▶ Vocabulary and Phrases

1. stringent [ˋstrɪndʒənt]
adj. （規定等）嚴格的；嚴厲的

The government imposed stringent controls on air pollution.
政府對空氣污染施加嚴格的管制。

2. revolutionary [͵rɛvəˋluʃə͵nɛrɪ] *adj.*
革命性的；完全創新的

This revolutionary product is very easy to use.
這項創新的產品很容易使用。

3. patented [ˋpætn̩tɪd] *adj.* 取得專利的

The toothbrush's design is patented.
那支牙刷的設計有取得專利。

4. microprocessor [͵maɪkroˋprɑ͵sɛsɚ] *n.* 微處理器

Tim decided to upgrade his computer with a faster microprocessor.
提姆決定裝一個更快的微處理器來使電腦升級。

5. activate [ˋæktə͵vet] *v.* 啟動

Susan activated the credit card as soon as she received it.
蘇珊一收到那張信用卡就開卡了。

6. hygiene [ˋhaɪ͵dʒin] *n.* 衛生

Many illnesses are caused by poor hygiene.
許多疾病都是起因於衛生不良。

7. plumber [ˋplʌmɚ] *n.* 鉛管工

The toilet wasn't working, so they called a plumber.
馬桶壞了，所以他們打電話叫水管工人來。

8. consumption [kənˋsʌmpʃən] *n.* 消耗

The consumption of too much junk food can lead to health problems.
吃太多垃圾食物會導致健康問題。

9. adapter [əˋdæptɚ] *n.* 接頭；轉接器
（= adaptor）

Terence needed an adapter to use the laptop computer.
泰倫斯需要一個轉接器才能使用那台筆記型電腦。

10. theft-proof [ˋθɛft͵pruf] *adj.* 防盜的

The car featured an alarm system and theft-proof door locks.
那款車的特色是有警報系統與防盜門鎖。

Attachment #2

HANDSOFF™ AND HANDY COMPARISON

	HANDSOFF™	HANDY
Model	• HO-1A	• HD-4C
Power Source	• 1 x 12 volt	• 2 x AA
Battery Life	• Nine to 12 months (at 150 uses / day)	• Three months (at 110 uses / day)
Installation	• Easy four-step process • No tools • Adapter does not damage faucet threads	• More difficult process • Requires screwdriver • Mounting adapter harms faucet threads
Sensing Technology	• Ultrasound • Senses **motion**[1] • Automatically adjusts sensing range to size of sink	• **Infrared**[2] • Senses presence of heat • Often activated by people walking by
Operation	• Simply put hand or object under unit to activate — unit automatically stops when hand or object is removed • User can manually stop faucet	• Must pass hand or object again to deactivate unit • No way to manually stop faucet in case of **malfunction**[3]
Design	• Attractive design **blends**[4] with American and European faucets	• Design looks **out of place**[5] with American and European faucets
Model Option(s)	• Standard unit • Theft-proof unit (locks into place)	• One standard unit only

中譯 附件二

HANDSOFF™ 與 HANDY 的比較

	HANDSOFF™	HANDY
型號	• HO-1A	• HD-4C
電源	• 一顆十二伏特的電池	• 兩顆 AA 電池
電池壽命	• 九到十二個月（一天使用一百五十次）	• 三個月（一天使用一百一十次）
安裝	• 簡單的四個步驟 • 不需要工具 • 轉接器不會傷到水龍頭螺紋	• 過程比較困難 • 需要用螺絲起子 • 安裝轉接器會傷到水龍頭螺紋
感應技術	• 超音波 • 感應（物體的）移動 • 自動將感應範圍調整為水槽的大小	• 紅外線 • 感應釋放的熱度 • 通常有人走過便會被啟動
使用	• 只要把手或物體放在裝置的下方就可開啟 — 手或物體移開時，裝置便自動關閉 • 使用者可手動關閉水龍頭	• 必需再把手或物體伸過去才能關閉裝置 • 如果發生故障，也無法手動關閉水龍頭
設計	• 有吸引力的設計與美國和歐洲的水龍頭很相配	• 設計與美國和歐洲的水龍頭看起來不搭配
款式選擇	• 一般型 • 防盜型（有上鎖）	• 只有一般型

► Vocabulary and Phrases

1. **motion** [ˈmoʃən] *n.* 運動；移動

   The scientists tracked the motion of the planet.
   那些科學家們追蹤那顆行星的移動。

2. **infrared** [ɪnfrəˈrɛd] *n.* 紅外線（亦可當 *adj.*）

   The infrared goggles allowed the soldiers to see people in the dark.
   紅外線的護目鏡讓士兵們可以在黑暗中看到人。

3. **malfunction** [mælˈfʌŋkʃən] *n.* 故障；機能不全

   If there is a malfunction, the company will replace the product.
   如果有故障，該公司將會更換那個產品。

4. **blend** [blɛnd] *v.* 協調；相稱

   The chair really blends with the rest of the furniture in the room.
   那張椅子真的和房間裡其他的傢俱很相配。

5. **out of place** 不相稱的；不合適的

   The extremely tall man looked out of place while playing with kids.
   那個非常高的男子在和小孩們玩耍時看起來很突兀。

Unit Two
Product Development
產品開發

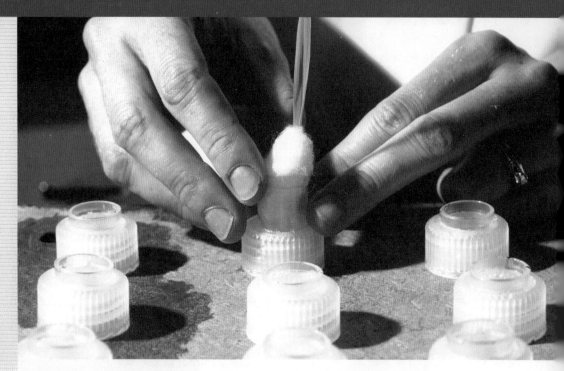

開發新產品的工作千頭萬緒，模具開好之後一般要經過試模、修模幾次才能最後定案，產品設計也可能要反覆修改才能通過品管檢驗測試。因此一般新產品開發至量產出貨的時間，都會比預期的還要長。

✔ Learning Goals

1. 新產品的開發應有專人負責，協調公司各部門及協力廠商之間的合作

2. 在決定要和哪家 **OEM** 廠商合作開發新產品前，需將配合條件講清楚，以使合作順利進行

Before We Start 實戰商務寫作句型

Topic: Talking about a New Product 談論新產品

◆ They expect / hope / intend to get [the production / the inspection / construction . . .] started / begun / commenced on / in / at / by (time).
他們預計／希望／打算在（時間）（之前）開始 [生產／檢查／建設工程……]。

◆ They will likely (probably) / It looks like they will have samples available / accessible / ready / prepared for [testing / evaluation / demonstration . . .] (at a certain time).
他們可能（在某時）會有樣品準備好供 [測試／評估／展示……]。

◆ Assuming (that) / Provided (that) the testing goes smoothly / runs well / goes according to plan, they are aiming for / targeting / seeking a test run (at a certain time).
假設測試進行順利／照計畫進行，他們的目標是（在某時）作試產。

◆ That's the [official / ideal / tentative . . .] schedule, which doesn't take the following factors into consideration / account . . .
那是在不考慮下列因素下 [官方的／理想的／暫時的……] 時間表……

◆ My fear / concern / worry is that S. + V. . . .
我擔心……

◆ [A pilot run / Shipping / An inspection . . .] is scheduled / slated / planned for (a certain time).
預定／計畫在（某時）[作試產／出貨／檢驗……]。

◆ The [prepayment / production delay / poor quality . . .] is a pivotal / crucial / critical / key issue right now.
[提前付款／生產延遲／品質不良……] 是目前很重要的問題。

A. Project Update 案子最新進展

Richard 向總公司報告工廠產品開發預計的進度及面臨的難題，並建議要統一窗口，以免影響開發的進程。

To: Scott Anderson
From: Richard Barlow
Date: 6/25
Subject: HandsOff™

Scott,

Mica expects to get the injection molds started next week. They will likely have samples available for testing in September. Assuming the testing goes smoothly, they are aiming for a test run in mid-December. That's the official schedule, which doesn't **take** the following factors **into consideration**:[1]

1. Mica says the equipment needed to make the PCBs will take an **indeterminate**[2] time period to get **debugged**[3] and ready for production.

2. Apparently, Mica HQ gave the US$12 quote without consulting Mica Taiwan, who received a quote of US$9 for only the board. Mica Taiwan is **befuddled**[4] by the quote of US$12 for the whole unit, assembled and packaged.

3. Mica Taiwan's fear is that after they place the tooling orders, these issues will become big enough to cancel the program, and they'll **be stuck with**[5] half-finished tooling.

COMMENT: My concern about the HandsOff™ project is that there seem to be too many players with no one person coordinating. We have Maples US, MAT, Mica Taiwan, Mica HQ, and the original designer of the product, ISR. We don't **foresee**[6] a plan that more than two or three of these parties will agree upon. Who's in charge?

Please send me any feedback you have.

Cheers,
Richard

中譯 收件人：史考特・安德森
寄件人：理查・巴洛
日期：六月二十五號
主旨：HandsOff™

史考特：

　　Mica 工廠預計在下星期開始開射出模。他們可能在九月會有樣品準備好供測試。假設測試進行順利，他們的目標是在十二月中作試產。那是在不考慮下列因素下官方的時間表：

1. Mica 工廠表示需要用來製造印刷電路板的設備不知道要花多久的時間來除錯並作好生產準備。

2. 顯然地 Mica 總公司在未諮詢其台灣分公司的情況下，就報價十二元美金。Mica 台灣分公司收到的報價光板子就要九塊美金，因此他們對整個產品連同組裝和包裝只報價十二元美金感到困惑不解。

3. Mica 台灣分公司擔心在他們下了模具的訂單後，這些問題會嚴重到導致整個案子被取消，而他們也將陷入模具僅完成一半的困境。

評論：我對於 HandsOff™ 這個案子擔心的地方在於太多人涉入卻沒有人來協調。我們有 Maples 美國總公司和台灣分公司、Mica 台灣分公司及其總部，還有這項產品的原始設計者 ISR。我們預見沒有一項計畫是其中兩、三方都會達成共識的。誰來主導？

如果你有任何意見請回信給我。

工作愉快

理查

▶ Vocabulary and Phrases

1. take sth into consideration 考慮到某事物

We have to take shipping costs into consideration when quoting a product.
我們做產品報價時必需考慮到貨運成本。

2. indeterminate [ˌɪndɪˈtɜmənɪt] *adj.*
難以確定的；限期不定的

He'll be away on a business trip for an indeterminate length of time.
他將要去出差，但不確定要去多久。

3. debug [diˈbʌg] *v.*
移去（電腦程式中的）錯誤；排除故障

The computer has a virus and needs to be debugged.
那台電腦有病毒需要除錯。

4. befuddled [bɪˈfʌdl̩d] *adj.* 迷惑不解的

The police are befuddled by how the money was stolen.
警方對於那些錢是如何被偷的感到迷惑不解。

5. be stuck with 無法擺脫；被……困住

The garbage man doesn't come until Monday, so I'm stuck with the trash.
收垃圾的要到星期一才來，所以我無法擺脫這些垃圾。

6. foresee [forˈsi] *v.* 預見；預知

The investors foresaw a lot of growth in the Asian market.
那些投資者預見在亞洲市場會有很大的成長。

Language Corner

1. Talking about Time Frames

It takes an indeterminate time period to V.
An indefinite amount of time is required to V.
We're not sure how long it's going to take to V.

For example:

1. An indefinite amount of time is required to assemble the products.

2. We're not sure how long it's going to take to complete the required repairs.

2. Grammar: There seem(s) to be . . .

使用該句型時，要注意如果 to be 後面接單數名詞要用 seems；接複數名詞則用複數形 seem。試比較下例：

✦ There <u>seems</u> to be a mistake.

✦ There <u>seem</u> to be a lot of mistakes.

Richard 向總公司報告模具已完成、工廠正要進行試產及協力廠商要求最低訂購數量的保證。

To: Scott Anderson
From: Richard Barlow
Date: 9/13
Subject: Product Update

Dear Scott,

We met with Mica today and discussed the following HandsOff™ issues:

1. All tooling sourced by Mica is virtually completed (all at least passed second shots), according to Mr. Yang.

2. A pilot run* of 25 units is scheduled for the end of October.

3. Regarding the PCBs, Yang says that the vendors want a minimum order of 100,000 pieces. That would be a US$900,000 order, which Maples must guarantee to Mica before he'll proceed. He **claims**[1] they can save us a lot of money by installing the equipment and making the boards themselves, but that would **put us out**[2] who-knows-how-many months.

4. This wasn't the first time we heard his point about the PCB vendors requesting the money up front for an order. Did you have the same problem with the companies who gave you quotes? If not, I think it would be worthwhile getting details of their expected terms, because the prepayment for a 100,000-board order is a **pivotal**[3] issue right now for Yang.

I'm glad to be able to **wind up**[4] with a bit of good news. Yang said he put together two units to test them for **leakage**[5] and they looked good.

Take care,
Richard

中譯 收件人：史考特・安德森
寄件人：理查・巴洛
日期：九月十三號
主旨：產品最新進度

親愛的史考特：

我們今天和 Mica 碰面討論下列有關 HandsOff™ 的問題：

1. 楊先生說所有 Mica 外包的模具差不多都已完成（全部至少通過第二次的試模）。

2. 預計十月底要試產二十五組的產品。

3. 有關印刷電路板的部分，楊先生説供應商要求的最低訂購數量為十萬件，那會是九十萬元美金的訂單。Maples 需先向 Mica 作保證，他才會繼續進行下去。他宣稱他們自己安裝設備來生產板子可以替我們省下很多錢，但那會造成我們的不便，誰曉得要等多久？

4. 這並不是我們第一次聽他説：「印刷電路板的供應商要求下單要先預付費用」。你跟向你報價的那些公司有碰到相同的問題嗎？如果沒有，我想值得去了解他們期望交易條件的細節，因為對楊先生來説要預付十萬塊板子訂單的費用是目前很重要的問題。

我很高興最後能以一點點好消息來作結。楊先生説他組裝了兩塊板子並測試是否有漏水，而它們看起來都很好。

保重

理查

Vocabulary and Phrases

1. claim [klem] *v.* 聲稱；主張

The caller claimed to work for the financial company, but I didn't trust her.

打電話來的人聲稱是替那家財務公司工作，但我不相信她。

2. put sb out 讓某人覺得不方便

I hope our early arrival didn't put you out.

我希望我們提早抵達沒有造成你的不便。

3. pivotal [ˋpɪvət!] *adj.* 重要的；關鍵的

The speech turned out to be a pivotal moment in the politician's career.

那場演講最後竟成了那位政治人物生涯中最關鍵的時刻。

4. wind up 使結束

The manager wound up the meeting by congratulating his top salesman.

那位經理藉由恭賀他最優秀的業務員來結束會議。

5. leakage [ˋlikɪdʒ] *n.* 漏水；滲漏

There was water on the kitchen floor, so he checked the refrigerator for leakage.

廚房的地板有水，所以他查看冰箱是否有漏水。

Biz Focus

★ **pilot run** 少量的試產

新產品要量產前，工廠會作少量的試產，以評估生產流程是否需調整，試產的數量會比 test run 多且至少要做 25 個，因 25 個樣品所取得的實驗數據用統計學來做分析才有意義。少量試產的產品也是訂單的一部分，通過品管檢驗即可出貨。

Language Corner

Talking about an Inconvenience

That would put us out who-knows-how-many months.

You Might also Say . . .

- That would delay us for a yet-to-be-determined amount of time.
- That would inconvenience us for an undetermined stretch of time.
- That would prevent us from progressing for an indefinite time period.

Unit Three
Orders 訂單

下給工廠的訂單內容需包含訂單號碼、品名、產品代號、數量、單價、總金額、交貨期、付款方式、收件人姓名及地址、報關行（或貨運公司）資料、運送方式及品質標準等。

 Learning Goals

1. 要收到印有客戶公司抬頭的訂購單（Purchase Order / PO）才算正式下單

2. 工廠應仔細核對訂單資料，簽名確認後再回傳給客戶

3. 小額交易以電匯方式來付款會比用信用狀來得方便且省錢

Before We Start 實戰商務寫作句型

Topic: Making a Purchase Order 寫訂購單

- ◆ I would like to clarify / make clear / clear up a few points before we proceed / move on / advance / go ahead.
  在我們繼續進行下去之前，我想澄清一些疑點。

- ◆ Your proposed / suggested [payment terms / commission request / deadline . . .] sound(s) fine / is (are) acceptable / is (are) satisfactory.
  你所提議的 [付款條件 / 佣金要求 / 最後期限……] 聽起來沒問題 / 是可以接受的 / 是令人滿意的。

- ◆ A [wire transfer / check / letter of credit . . .] would be the most sensible / most practical / best approach.
  [電匯 / 支票 / 信用狀……] 是最明智 / 最實際 / 最好的方式。

- ◆ Please advise / inform / tell us if you have a forwarder of choice / you'd prefer, and to whom the [shipment / package / parcel . . .] should be addressed / sent / delivered.
  請告知我們你是否有偏好的貨運公司，以及 [這批貨 / 這個包裹……] 該寄給誰。

- ◆ (Company A) has this day bought / purchased the undermentioned / following goods from (company B) to be delivered in good order subject to the terms and conditions [stated / listed / detailed . . .] hereunder / below, unless otherwise specified . . .
  （A 公司）於今日向（B 公司）購買下列（提及的）物品，除非另外註明，否則這些貨物都需依照下面 [陳述 / 條列 / 詳述……] 的條款完好地運達……

- ◆ Please [acknowledge / confirm / verify . . .] receipt and acceptance / approval of this order as soon as is convenient / at your earliest convenience by returning the copy duly signed.
  請以適當方式簽名並回傳副本，儘早 [告知 / 確認……] 收到並接受這份訂單。

A. Confirming Order Details 確認訂單細節

客戶準備要購買模具，Richard 特地寫信向對方澄清一些細節，並要求其正式下單。

To: William Kay
From: Richard Barlow
Date: 3/26
Subject: Project L-0202

Mr. Kay,

Regarding Project L-0202, I would like to clarify a few points before we proceed:

A. As you know, the notes on the drawing we have are in Dutch. Would you please translate them into English?

B. In response to your request about getting a **refund**[1] on the tooling down payment if parts are not approved: typically, if there are problems with the first shots, the toolmaker will **modify**[2] the tool as many times as necessary. Of course, if the toolmaker doesn't deliver acceptable tooling for its own reasons, we will get the money back.

C. Your proposed payment terms sound fine. For this rather small amount, a wire transfer would be the most **sensible**[3] approach.

D. Please advise us if you have a **forwarder**[4] **of choice**[5] in Taiwan, and to whom the shipment should be addressed.

E. Please send us a purchase order today, including the final payment terms as well as the names and addresses of the forwarders and brokers.

Thanks for all your assistance on the project.

Take care,
Richard

中譯　收件人：威廉‧凱

　　　寄件人：理查‧巴洛

　　　日期：三月二十六號

　　　主旨：專案 L-0202

凱先生：

關於專案 L-0202，在我們繼續進行下去之前，我想澄清一些疑點：

　　A. 如你所知，我們拿到的圖面上的註記都是荷蘭文。可以請你把它們翻成英文嗎？

　　B. 回應你提出有關如果零件未被驗收就要退還模具頭期款的要求：通常第一次試模如果有問題，模

具廠商會視情況所需一直修模。當然如果模具廠商因為自身的因素無法交出可以接受的模具，我們將會把錢要回來。

C. 你所提議的付款條件聽起來沒問題。像金額這麼小的款項，電匯是最明智的方式。

D. 請告知我們你在台灣是否有偏好的貨運公司，以及這批貨該寄給誰。

E. 請於今天寄給我們訂購單，其中需包含確定的付款條件以及貨運公司與代理人的姓名、地址。

感謝你對這個案子所提供的所有協助。

保重
理查

Vocabulary and Phrases

1. **refund** [ˋriˏfʌnd] *n.* 退還；退款

   The toy was broken, so his father took it back to the store to get a refund.
   那個玩具壞掉了，所以他爸爸拿回店裡要求退錢。

2. **modify** [ˋmɑdəˏfaɪ] *v.* 更改；修改

   He modified the design to make the product easier to use.
   他修改了設計讓產品更容易使用。

3. **sensible** [ˋsɛnsəb!] *adj.* 明智的；合情理的

   Buying a car that didn't use up a lot of gasoline was a sensible decision.
   買一台不會耗很多油的車是個明智的決定。

4. **forwarder** [ˋfɔrwɚdɚ] *n.* 運輸業者；運送者

   Our forwarder failed to deliver the goods to their destination.
   我們的貨運公司沒有把貨物送到目的地。

5. **of choice** 偏好的；屬意的

   The computer program was the tool of choice for graphic designers.
   那個電腦程式是平面設計師所偏好的工具。

Language Corner

Guaranteeing Work Quality

The toolmaker will modify the tool as many times as necessary.

You Might also Say . . .

- The toolmaker will make changes to the tool as many times as needed.
- The toolmaker will alter the tool as many times as called for.
- The toolmaker will work on the tool until it gets approved.

AECO 替 Digitron 公司向供應商 Well Sing 採購電線，並請 FETS 負責作出貨品管。下列為 AECO 所下的訂單。

AECO LTD.

US Office
3906 Cabrillo Street, Suite 504
Long Beach, CA 90021
213-831-0003
cust_serv@aecoltd.com

PURCHASE ORDER

NUMBER: 002 / 02   *DATE: Feb. 18*

AECO Ltd. has this day bought the **undermentioned**[1] goods from Well Sing Industrial Co., Ltd. to be delivered in good order **subject to**[2] the terms and conditions stated **hereunder**,[3] unless otherwise specified:

DESCRIPTION	QUANTITY	UNIT PRICE	AMOUNT
P / N 01011 Straight Cable, as last delivered	1,000 PCS*	US$1.30	US$1,300.00

SHIPPING DOCUMENTS: Please list Digitron Research Corp. as the buyer and **consignee**.[4]

DESTINATION: Digitron Research Corp.
     c / o* Jake Prosper Brokers
     5578 Lawrence Blvd.
     Eugene, OR 78323
     Attn: Tina Charles

DELIVERY: Two weeks or ASAP

PACKING: For air freight — ship freight collect*

QUALITY: Same as last PO (thru Hi-tronics)

REMARKS:[5] Far East Technical Services to be AECO's agent for all technical questions, quality inspection, communication, and shipping information

Please **acknowledge**[6] receipt and acceptance of this order as soon as is convenient by returning the copy **duly**[7] signed.

Confirmed and Accepted by   Buyer's Signature

  David Wu   *Ken Smith*
_____   _____
   (Well Sing)   (AECO)

中譯　AECO LTD.

90021 加州長灘市

卡布里奧街 3906 號 504 室

美國辦公室

213-831-0003

cust_serv@aecoltd.com

訂購單

編號：*002 / 02*　　日期：二月十八號

AECO Ltd. 於今日向 Well Sing Industrial Co., Ltd. 購買下列提及的物品，除非另外註明，否則這些貨物都需依照下面陳述的條款完好地運達：

描述	數量	單價	總計
產品編號 01011 直電線 — 同上一批的出貨	一千件	1.3 美元	1,300 美元

出貨文件：請將 Digitron Research Corp. 列為買
主及收件人。

目的地：Digitron Research Corp.
　　　　由 Jake Prosper Brokers 轉交
　　　　請寄到
　　　　78323 美國奧勒崗州尤金市
　　　　羅倫斯大道 5578 號
　　　　收件人：蒂娜‧查爾斯

交貨期：兩個星期或儘快

包裝：空運及海運由買方到付

品質：同上一批訂購單（透過 Hi-tronics 採購的）

註釋：Far East Technical Services 為 AECO 的
　　　代理人，負責所有技術問題、品質檢驗、溝通
　　　以及出貨資訊

請以適當方式簽名並回傳副本，儘早告知收到並接受
這份訂單。

經確認及接受　　　買主簽名
　　吳大衛　　　　　肯‧史密斯
（Well Sing）　　　（AECO）

Vocabulary and Phrases

1. **undermentioned** [ˌʌndɚˈmɛnʃənd] *adj.* 下面提到的
   Please take a look at the undermentioned proposal.
   請看一下下面提到的計畫書。

2. **subject to** 受制於
   All US contracts are subject to the laws of that country.
   所有美國的合約都需遵照該國的法律。

3. **hereunder** [ˌhɪrˈʌndɚ] *adv.* 在下（文）；據此
   Please supply me with everything listed hereunder.
   請提供我下面列出的每樣物品。

4. **consignee** [ˌkɑnˌsaɪˈni] *n.* 收件（貨）人
   The consignee expected the shipment by December 10.
   收件人期望能在十二月十號前收到那批貨。

5. **remark** [rɪˈmɑrk] *n.* 評論；意見
   Shannon didn't appreciate her colleague's remarks about her performance.
   夏儂並不感激她同事對她表現的評論。

6. **acknowledge** [əkˈnɑlɪdʒ] *v.* 告知收到（信件等）
   I called him to acknowledge receipt of the note.
   我打電話告訴他我已經收到那張便條。

7. **duly** [ˈdjulɪ] *adv.* 適當地
   All packages must be duly labeled, or they will not be delivered.
   所有的包裹都必需被適當地貼上標籤，否則無法被遞送。

Biz Focus

★ PCS 件數（為 pieces 的簡寫，單數用 PC）

★ c / o 由……轉交（= care of / in care of）

★ freight collect 運費（由買方）到付
　 freight prepaid 則是指「運費先（已）付」。

Unit Four
Packaging 包裝

客戶提供包裝設計圖（artwork）及底片（film）請 OEM 代工廠製作產品的包裝，代工廠收到資料後需替客戶校稿、核對包裝和實際產品的尺寸等，務求在生產前將所有問題找出來並做更正。

✅ Learning Goals

1. 彩色或反白內容的更改需由客戶端提供新的底片，白底黑字內容的更改工廠則可自行修改

2. 基於環保的訴求，避免過度包裝產品並儘量使用回收材料

3. 產品包裝應包括品名、型號、商標、產地、UPC 條碼等資料

Before We Start Shipping Marks 出貨嘜頭

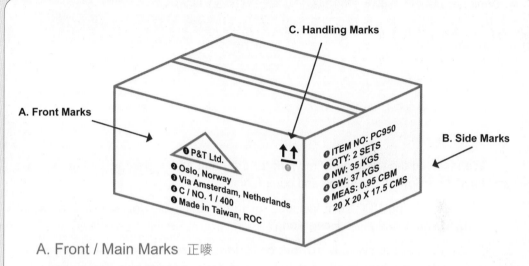

C. Handling Marks

A. Front Marks

❶ P&T Ltd.
❷ Oslo, Norway
❸ Via Amsterdam, Netherlands
❹ C / NO. 1 / 400
❺ Made in Taiwan, ROC

❶ ITEM NO: PC950
❷ QTY: 2 SETS
❸ NW: 35 KGS
❹ GW: 37 KGS
❺ MEAS: 0.95 CBM
　 20 X 20 X 17.5 CMS

B. Side Marks

A. Front / Main Marks 正嘜

❶ 進（出）口商的縮寫
❷ 目的地名稱
❸ 表示該批為轉運貨物，via 後面接「轉運點」
❹ 紙箱流水號，亦可寫成「C / NO. 1 OF 400（四百箱中的第一箱）」
❺ 產地

B. Side Marks 側嘜

❶ 貨號
❷ 數量（= quantity）
❸ 淨重（= net weight）
❹ 毛重（= gross weight）
❺ 尺寸（= measurements）

C. Handling Marks 貨運標誌

❶ This Side Up 此面朝上

其他常見的貨運標誌如下：

| **Fragile** | **Handle with Care** | **Keep Dry** | **Do Not Stack** | **Flammable** |
| 易碎物品 | 小心輕放 | 保持乾燥 | 不要疊放 | 易燃物 |

註：貨運標誌亦可說成 pictorial markings for handling goods。

Richard 在收到包裝設計圖後，寫信與客戶釐清一些包裝上的疑點。

To: Ray Franklin
From: Richard Barlow
Date: March 20
Subject: Comments about Artwork

Dear Mr. Franklin,

We received the package of artwork for the hand tool yesterday. I've reviewed it and have the following questions and comments:

1. There was no UPC* film included for the product. Did you **scrap**[1] the use of UPCs on packages in Hong Kong?

2. We did not receive any artwork or film for an instruction book or any other **inserts**,[2] like a **warranty**[3] card or a list of **depot services**[4] in this package. Do you know if more artwork and film for this product is on the way?

3. The model number on the front and back of the insert card is wrong. The printing on the back of the card is black, so we can correct the model number there. However, the one on the front is printed in **white reverse type**,[5] which means we need four new sheets of film to correct it. Since the matter is **pressing**,[6] I suggest we put a sticker over the incorrect model number on the front of the card in the meantime.

4. We didn't receive any shipping marks or artwork for the shipping cartons. Please tell us what we should print on them.

I look forward to your response.

Best regards,
Richard Barlow

中譯　收件人：雷・法蘭克林
　　　寄件人：理查・巴洛
　　　日期：三月二十號
　　　主旨：對包裝設計圖的意見

親愛的法蘭克林先生：

　　我們昨天收到手動工具包裝設計圖的包裹。我看過之後有下列的問題和意見：

　　1. 裡面沒有針對該產品通用代碼的底片。在香港，你們的包裝不用通用產品代碼嗎？

　　2. 在包裹內我們沒有收到任何使用手冊的設計圖或底片，或其他內附卡片像保證卡、維修站服務清單等。您知道是否有更多這項產品的設計圖和底片正要送過來嗎？

　　3. 插卡正反面的型號有誤。插卡背面是用黑色印刷，所以我們可以修改型號。然而插卡正面是

採反白印刷，這表示我們需要四張新的底片才能作修改。既然這件事很急迫，我建議在這期間先在插卡正面錯誤的型號上貼標籤。

4. 我們沒有收到任何出貨紙箱的嘜頭或設計圖。請告訴我們上面應該要印什麼。

我期待收到您的回信。

謹上
理查‧巴洛

Vocabulary and Phrases

1. scrap [skræp] *v.* 丟棄；報廢

If we find that the costs are too high, we might just scrap the trip completely.
如果我們發現費用太高，我們有可能捨棄整個行程。

2. insert [`ɪnsɝt] *n.* 插頁；插入物

There are a lot of advertising inserts in the newspaper during the holidays.
假日期間報紙內都會有大量的夾頁廣告。

3. warranty [`wɔrəntɪ] *n.* 擔保；保證書

There was a three-year warranty on the new car he bought.
他買的新車有三年保固。

4. depot service （此指）維修站服務

Amy wanted the technician to come to her house to fix the computer, but he said they only offered certain depot services.
艾美想請技術人員來她家修電腦，但他說他們只提供一些維修站服務。

5. white reverse type 反白

The Web page's background was dark, so its designer used white reverse type.
這個網頁的背景是黑的，所以它的設計者用了反白字體。

6. pressing [`prɛsɪŋ] *adj.* 急迫的；緊急的

The company's executive scheduled a meeting to talk about a pressing issue.
公司主管安排會議討論一個緊急的議題。

Biz Focus

★ **UPC** 通用產品代碼（= Universal Product Code）

UPC 源自於美、加等北美洲地區；EAN 歐洲商品條碼（= European Article Number）則源自於歐洲，目前這兩套商品代碼都被廣為使用。

Language Corner

Solving an Urgent Matter

Since the matter is pressing,	
Considering this business is urgent,	} I suggest (that) S. (should) V. . . .
Because this problem needs to be solved right away,	

For example:

1. Considering this business is urgent, I suggest we ship the package via express mail.

2. Because this problem needs to be solved right away, I suggest that we call a meeting for this afternoon.

B. Printing Requirements 印刷要求

Richard 向總公司報告一定要將產地名稱印在產品上或貼上有產地字樣的小標籤，才能在台灣通關出貨。

TO: James McInerney
FROM: Richard Barlow
DATE: 10/4
SUBJECT: "Made in Taiwan" Stickers

James,

Tyokai refused to have "Made in Taiwan" printed on their products because Japanese consumers don't like stuff that's made in Taiwan. We lightly attached a small "Made in Taiwan" sticker (with weak **adhesive**)[1] to each unit — leaving one end of the sticker **unstuck**[2] to the product for easy removal. Tyokai then removed all the stickers in Japan.

The only other example I can remember is the old Company BD C-1, which didn't have a "Made in Taiwan" label on the unit, but no one ever noticed. Taiwan **Customs**[3] just **spot-checks**[4] for it, so it's possible to ship products without it. If it is caught, however, the shipment is **retained**[5] by customs and rework is required. Not a good risk to take, I think.

Take care,
Richard

中譯

收件人：詹姆士・麥肯納利
寄件人：理查・巴洛
日期：十月四號
主旨：「台灣製造」標籤

詹姆士：

Tyokai 公司拒絕在他們的產品上印上「台灣製造」，因為日本的消費者不是很喜歡台灣製的產品。我們稍微貼了一個「台灣製造」的小標籤（用不是很黏的黏膠）在每一件產品上 — 標籤有一邊沒有黏住產品以方便撕掉。之後 Tyokai 在日本再將所有的標籤拿掉。

我記得唯一不同的例子是 BD 公司的舊產品 C-1，他們的產品上並沒有「台灣製造」的標籤，但從未有人注意到。台灣海關只是抽樣檢查，所以沒有標籤就出貨是有可能的。但如果被抓到，貨運會被海關扣留並需要重工。我認為沒有必要冒這個險。

保重
理查

Vocabulary and Phrases

1. **adhesive** [əd`hisɪv] *n.* 黏著劑
   We used very strong adhesive to reconnect the leg of the table.
   我們用了非常強的黏著劑來重新連接桌腳。

2. **unstuck** [ʌn`stʌk] *adj.* 未黏住的；脫落的
   The sticker on his hat became unstuck when it rained.
   下雨時，他帽子上的標籤變不黏了。

3. **customs** [`kʌstəmz] *n.* 海關
   The shipment was delayed for nine days because customs wanted to know the origin of the materials.
   那批貨遲到九天，因為海關要知道那些材料的來源。

4. **spot-check** [`spɑt͵tʃɛk] *v.* 進行抽查；做抽樣調查
   The inspector spot-checked samples that the factory produced.
   驗貨員對這間工廠生產的樣品進行抽查。

5. **retain** [rɪ`ten] *v.* 留住；保持
   The boxes of goods were retained at the border for further inspection.
   那幾箱貨物被留在邊境以做進一步的檢查。

6. **recyclable** [rɪ`saɪkləb!] *n.*
   可回收利用的材料（作此義解時，常用複數）
   The shopping bags were made out of recyclab
   那些購物袋都是用可回收利用的材料做的。

C. Packaging Material 包裝材料

加拿大政府要求進口產品的包裝要用一定比率的回收材料，Richard 乃向總公司呈報包裝材料種類及如何因應。

○ ○ ○

TO: Curtis Poppell
FROM: Richard Barlow
DATE: 11/20
SUBJECT: Cardboard

Curtis,

The Canadian government has requested that we increase our use of **recyclables**[6] and reduce the weight of our packages.

All the cardboard we use for display boxes is 100 percent recycled material. But the master cartons we now use are made from **virgin paper**.[7] I don't think this is going to be a problem though — we should be able to make the change fairly easily. I'll let you know tomorrow.

In addition, we can have Company #2B switch to the Company #4T-size display box. This will reduce the **overall**[8] package weight by 23.5 percent. We can think of nothing else, unless we **eliminate**[9] the dust caps. We've already checked with Mica for ideas and they had nothing to offer. If you have any suggestions, don't hesitate to share them.

Cheers,
eom

中譯

收信人：柯提斯・帕貝爾
寄件人：理查・巴洛
日期：十一月二十號
主旨：硬紙板

柯提斯：

加拿大政府要求我們多使用可回收利用的材料並減少我們包裝的重量。

我們的展示盒所使用的所有硬紙板百分之百都是回收的材質。不過我們目前使用的外箱是用原生紙做的。但我想這不成問題 — 我們應該可以很輕易地作更換。我明天會再告訴你。

此外，我們可以讓 #2B 公司改用 #4T 公司尺寸的展示盒。這將可以減少百分之二十三點五的總包裝重量。除非我們拿掉防塵蓋，否則我們想不出別的方法了。我們已經問過 Mica 的想法，他們沒有任何意見。如果你有任何建議，請儘量與我們分享。

工作愉快
完畢

---

virgin paper 原生紙
This virgin paper is more expensive than recycled paper.
這種原生紙比回收紙貴。

overall [ˋovɚͺɔl] *adj.* 全部的；總的
We made a profit this month, but our overall financial situation is not good.
我們這個月有盈利，但整體的財務狀況還是不好。

eliminate [ɪˋlɪməͺnet] *v.* 除去；消除
The CEO decided to eliminate unnecessary business travel to save money.
執行長決定去除不必要的出差以省錢。

Language Corner

Advising about Risks

Not a good risk to take, I think.

You Might also Say . . .

- In my opinion, we shouldn't take that chance.
- I wouldn't take that gamble.
- It is too uncertain to pursue.

Unit Five
Samples 樣品

展示間的陳列、內部做實驗測試或提供給客戶做技術評估、銷售展示等都需要用到樣品，因此通常生產產品的數量會比訂單出貨的量還要多一些，以因應各種不同的樣品需求。

✅ Learning Goals

1. 供品管檢驗用的限度樣本，需於量產時一併收集

2. 應以正式書面提出樣本需求，並註明型號、數量、包裝、出貨日期及運送方式等

3. 提供給客戶的樣品，需經 **QC** 部門全檢才可出貨

4. 應提供客戶樣品的貨運資料，以方便其作追蹤

Before We Start 實戰商務寫作句型

Topic: Handling Sample Requests 處理樣品需求

- ◆ We asked your QC department to collect / gather / put together limit samples that do not meet the various aesthetic requirements.
  我們要求你們的品管部門收集不符合各種外觀需求的限度樣品。

- ◆ Judgment of classifications of some defects is sometimes [subjective / opinion-based / subject to personal preference . . .].
  判定某些缺陷的類別有時 [很主觀 / 見仁見智 / 易受個人偏好影響……]。

- ◆ I propose we use limit samples to provide a frame of reference / set of standards.
  我建議我們用限度樣品來提供參照標準。

- ◆ We have arranged for these samples to be delivered / sent / shipped out to you tomorrow.
  我們已安排明天把這些樣品寄給你們。

- ◆ I can't be sure / guarantee / know for sure that we will have a surplus of / extra / spare / excess units on top of the order quantity.
  我不確定除了訂單的數量外，我們是否還有多餘的產品。

- ◆ We want to be certain / positive that the samples we send to a current or potential customer are [flawless / perfect / impressive / of the best quality . . .].
  我們想確保寄給現有或潛在客戶的樣品是 [完美無瑕的 / 令人印象深刻的 / 最優質的……]。

A. Limit Samples 限度樣品

#1 Richard 敦促工廠儘快收集外觀有瑕疵的產品來做限度樣品。

TO: Gary Yang
FROM: Richard Barlow
DATE: 6/14
SUBJECT: Collecting Limit Samples

Gary,

When production began, we asked your QC department to collect limit samples that do not meet the various **aesthetic**[1] requirements of the Royal Molding products (color differences, molding marks, etc.). I believe Mr. Alexander will be expecting to review these limit samples during his visit.

Several months ago, we also requested that your QC department put together limit samples for Tyokai. They need limit samples of color **deviation**[2] and visual defects for their model and packaging, but we haven't received anything yet. Please notify me when the limit sample boards* will be available. We need to FedEx them in the next few days.

Thanks for your attention to these matters.

Cheers,
eom

中譯

#1

收件人：楊蓋瑞
寄件人：理查·巴洛
日期：六月十四號
主旨：收集限度樣品

蓋瑞：

開始生產時，我們要求你們的品管部門收集不符合各種 Royal Molding 公司產品外觀需求的限度樣品（色差、射出模痕跡等）。我相信亞歷山大先生期望他來訪時可以檢視這些限度樣品。

幾個月前我們也曾要求你們的品管部門替 Tyokai 公司收集限度樣品，他們需要有關他們產品和包裝上有色差及外表缺陷的限度樣品，但我們還沒收到任何東西。請告知我何時可拿到附有限度樣品的板子。我們需在未來幾天內把它們快遞出去。

感謝你對這些事情的處理。

工作愉快
完畢

Vocabulary and Phrases

1. **aesthetic** [ɛsˋθɛtɪk] *adj.*
   美的；美學的（aesthetics *n.*）

   The new cell phone is durable, but I wish its aesthetic qualities could be improved.
   這支新手機很耐用，但我希望它的外觀品質可以再改進。

2. **deviation** [ˌdivɪˋeʃən] *n.* 誤差；偏差

   Any deviation from what is normal should be reported to the manager.
   任何偏離標準的誤差都應向經理報告。

3. **classification** [ˌklæsəfəˋkeʃən] *n.* 分類；分級

   The government changed the animal's classification from endangered to protected.
   政府將那種動物的分類從瀕臨絕種改成保護類。

4. **subjective** [səbˋdʒɛktɪv] *adj.* 主觀的

   You can't measure the painting's beauty. It's completely subjective.
   你無法測量那幅畫的美麗。那完全是主觀的。

#2 Richard 向客戶提議用限度樣品作為產品外觀檢驗的標準。

TO: James Alexander
FROM: Richard Barlow
DATE: 6/15
SUBJECT: Determination of Visual Defects

Dear Mr. Alexander,

Since judgment of **classifications**[3] of some defects — such as color differences and minor visual defects — is sometimes **subjective**,[4] I propose we use limit samples to provide a **frame of reference**[5] for judging those defects.

The limit samples for aesthetics (packaging, printing, molding, colors, etc.) need to be collected from production. Our OEM supplier, Mica, has been collecting them. Hopefully, we can review them during your visit. If not, we'll send them to you as soon as we have some complete sets.

Take care,
Richard Barlow

中譯

#2
收件人：詹姆士‧亞歷山大
寄件人：理查‧巴洛
日期：六月十五號
主旨：外觀缺陷的判定

親愛的亞歷山大先生：

既然判定某些缺陷類別（如色差和輕微的外觀缺陷）有時很主觀，我建議我們用限度樣品作為評判那些瑕疵的準則。

針對外觀（包裝、印刷、塑膠射出、顏色等）的限度樣品需在生產時收集。我們的代工生產商 Mica 已經有在收集，但願我們可以在您來訪時檢視那些樣品。如果來不及的話，等我們一拿到幾套完整的樣品就會馬上寄給您。

保重
理查‧巴洛

5. **frame of reference** 參照標準；準則

Our employee handbook provides a frame of reference for dealing with this type of situation.
我們的員工手冊有提供處理這種情況的參照標準。

▶ **Biz Focus**

★ **limit sample board** 附有限度樣品的板子
限度樣品會依瑕疵的嚴重程度排列在板子上。

Insight

Trademarks as Verbs 可以當作動詞使用的商標名稱

Word	Definition
FedEx	Send by overnight mail
xerox	Make a copy
google	Search / find information online

For example:

1. Can you xerox 50 copies of this report for me?
2. Can you google that company's Web site?

B. Sample Requests 樣品需求

#1 Richard 寫信請代工工廠提供樣品，並交待出貨給客戶的運送方式及如何請款。

TO: Jack Liang
FROM: Richard Barlow
DATE: 8/4
SUBJECT: Company #2B Sales Samples

Mr. Liang,

This is to confirm our sample request for the following units:

1. 50 AB-1 and all packaging components, **unsealed**[1] clamshells;* either packaging is OK.

2. 10 CD-1, no packaging.

Tyokai needs these units before 8/12. They should be **posted**[2] air freight collect to:

H. Asahina
Tyokai Corporation
395 Miyazaki Boulevard, Suite 132
Tokyo, Japan

Please correspond with Eric Woods for a T / T* to pay for these samples. Our US head office tells me they will **disburse**[3] the funds within 30 days of shipment.

Thanks,
Richard

中譯

#1
收件人：梁傑克
寄件人：理查‧巴洛
日期：八月四號
主旨：#2B 公司的銷售展示樣品

梁先生：

這封信是為了確認我們以下產品的樣品需求：

1. 五十件 AB-1 和所有包裝材料，隻泡殼勿封起來；兩種包裝皆可。

2. 十件 CD-1，不需要包裝。

Tyokai 需在八月十二號以前拿到這些產品。請以到付方式將它們空運至：

日本東京市宮崎大道 395 號 132 室
Tyokai 公司
H. 朝比奈先生 收

請與艾瑞克‧伍茲聯絡用電匯支付這些樣品的事宜。我們美國總公司告訴我他們會在出貨後的三十天內付款。

謝謝
理查

Vocabulary and Phrases

1. **unsealed** [ʌnˋsild] *adj.* 未封口的
The unsealed packages were waiting to be taped up and shipped.
那些未封口的包裹尚待封裝和出貨。

2. **post** [post] *v.* 郵寄；投遞
I'm going to post the letter this afternoon.
我今天下午要寄出這封信。

3. **disburse** [dɪsˋbɝs] *v.* 支付
Your earnings will be disbursed to your bank account automatically.
你的工資將會自動轉到你的銀行戶頭。

4. **collective** [kəˋlɛktɪv] *adj.* 集體的；共同的
We talked it over, and our collective opinion is that we should hire Steven.
我們討論過這件事，而我們共同的想法是應該聘用史帝文。

5. **surplus** [ˋsɝpləs] *n.* 過剩；剩餘物
We didn't sell as much as usual last month, so we have a surplus in our warehouse.
上個月我們並沒有和平常賣的一樣多，所以我們倉庫還有剩餘的產品。

#2 Richard 懷疑客戶重覆提出兩次樣品需求，於是寫信確認。

TO: Kent Roach
FROM: Richard Barlow
DATE: Sept. 27
SUBJECT: Test Samples

Dear Mr. Roach,

We received your sample request today, but I'm confused by your reference to an earlier request for sales samples.

To the best of our **collective**[4] memories, the only sample request we have received from you was one asking that one carton of each model be sent for use by your QA department. We have arranged for these samples to be delivered to you tomorrow.

Please let me know if you require an additional carton of each model. I must mention, however, that our vendor tries to manufacture as few extra units as possible, and I can't be sure that we will have a **surplus**[5] of units **on top of**[6] the order quantity. If you need the extra cartons, I'll try my best to put them together.

Regards,
Richard Barlow

中譯

#2

收件人：肯特‧洛奇
寄件人：理查‧巴洛
日期：九月二十七號
主旨：測試樣品

親愛的洛奇先生：

我們今天接獲了你們樣品的需求申請，但我對於你提到你們先前提出的銷售樣品需求感到困惑。

就我們記憶所及，我們唯一收到的樣品需求申請，是要求每個型號各寄一箱供你們品保部門使用。我們已安排明天把這些樣品寄給你們。

請讓我知道你們是否還需要每個型號多寄一箱的樣品。但我必需説明我們的供應商儘可能不生產多餘的產品。我不確定除了訂單的數量外，我們是否還有多出的產品。如果你需要額外的量，我會儘量收集。

祝 商安
理查‧巴洛

6. on top of 除……之外

On top of your salary, you will receive a bonus.
除了你的薪水之外，你還會得到一筆獎金。

▶ **Biz Focus**

★ clamshell 雙泡殼（類似蛤殼的產品包裝）

★ T／T 電匯（= telegraphic transfer）

Language Corner

Recalling Something

✦ To the best of our collective memories, . . .
✦ As far as we can remember, . . .
✦ If we recall correctly, . . .
✦ If our memory serves us well, . . .

For example:

1. As far as we can remember, your order was sent out February 21.
2. If we recall correctly, we haven't had a meeting with you since June.

C. Sample Inspection 樣品檢驗

Richard 除轉達美國總公司的樣品需求外，特別交待代工生產商要讓 Maples 派駐在工廠的檢驗員驗過樣品後才可出貨。

TO: Yolanda Ho
FROM: Richard Barlow
DATE: 9/26
SUBJECT: Samples for Mr. Woods

Yolanda,

Mr. Woods needs the following samples sent to him **on the double**:[1]

36 — Company #3C CA-1 (with or without stickers)

24 — Company #4D TC-2

24 — Company #6P PH5577 (not produced yet)

I believe the CA-1s and TC-2s are available now. Please send them, and ship the other units as they become available.

You noted in a recent message that these samples were already in Hong Kong. But I don't think they were 100 percent inspected by Maples QC engineer in China. We want to be positive that the samples we send to a current or potential customer are **flawless**.[2]

Let me **reiterate**[3] our sample policy to **refresh your memory**.[4] All samples sent from the China factory must be first inspected and **readied**[5] by Harry, our engineer in your factory, before you ship them to our customers.

Thanks,
Richard

中譯　收件人：何尤蘭妲
　　　寄件人：理查‧巴洛
　　　日期：九月二十六號
　　　主旨：伍茲先生的樣品

尤蘭妲：

伍茲先生需要立即收到下列樣品：

三十六件 #3C 公司的 CA-1（附或不附標籤均可）

二十四件 #4D 公司的 TC-2

二十四件 #6P 公司的 PH5577（尚未生產）

我相信 CA-1 和 TC-2 的樣品都已經有了，請將它們寄出，其他產品等有樣品時再寄。

妳在最近的一封信息裡提到這些樣品已經在香港了。但我不認為它們有被 **Maples** 在大陸的品管工程師百分之百檢驗過。我們想確保寄給現有或潛在客戶的樣品是完美無瑕的。

讓我再次重申我們的樣品政策以喚醒妳的記憶。所有大陸工廠的樣品在寄給我們的客戶前，都需先讓我們派駐在你們工廠的工程師哈利檢驗過並做好出貨準備。

謝謝

理查

► Vocabulary and Phrases

1. on the double 立即

I need you to finish this report on the double.
我需要你立刻完成這份報告。

2. flawless [`flɔlɪs] *adj.* 無瑕疵的；完美的

The orchestra's performance was flawless.
那個管絃樂團的表演完美無瑕。

3. reiterate [ri`ɪtəˌret] *v.* 重申；反覆做

Samantha reiterated that all employees must arrive at eight a.m.
莎曼莎重申所有員工必需在早上八點到。

4. refresh one's memory 使恢復記憶；使想起

Before the speech, he refreshed his memory by reviewing his notes.
在演講前他藉由看筆記來喚醒自己的記憶。

5. ready [`rɛdɪ] *v.* 預備；使準備好

Sandy and Tom readied the boat for their upcoming sailing trip.
姍蒂和湯姆為他們即將到來的航行之旅準備好了船隻。

Language Corner

Clarifying in Different Tones

Let me reiterate our sample policy to refresh your memory.

> *You Might also Say . . .*
>
> ▪ I think there is a need for me to go over our sample policy again.
> ▪ You should be familiar with our sample policy by now.
> ▪ Apparently, you need to be reminded of our sample policy yet again.

D. Sample Shipment 樣品出貨

#1 逆滲透濾水器製造商向 **Richard** 回覆有收到他傳的樣品訂單，並告知出貨時間。

Primary Manufactured Systems, Inc.

FAX COVER SHEET

DATE SENT:	RECIPIENT:	COMPANY:
1/24	R.A. BARLOW	MAPLES CORPORATION

FAX NO.:	TRANSMITTED[1] BY:
886-4-2356-0800	FRANK HEMINGWAY

TOTAL PAGES IN THIS TRANSMISSION: 1

Thank you for your order of an R0-S1V system. We will ship the unit by February 3rd to ensure it arrives in Taiwan **with plenty of time to spare**.[2] With shipping charges, the total is US$292. We will also include a CD that explains the **reverse osmosis**[3] process, plus some **preliminary**[4] test data on the system from an independent test laboratory that can be used to prove its effectiveness in sales presentations. You will probably find the general FAQs regarding the R / O process helpful as well.

中譯

#1

Primary Manufactured Systems, Inc.

傳真封面

傳送日期：一月二十四號
收件者：R.A. 巴洛
公司：Maples 公司
傳真號碼：886-4-2356-0800
傳真者：法蘭克・海明威
總傳真頁數：一頁

感謝您訂購 RO-S1V 系統。我們將在二月三號以前寄出設備，以確保它有充裕的時間運抵台灣。含運費的價格總計是兩百九十二元美金。我們也將附上一張 CD 解釋逆滲透的過程，以及一些來自獨立測試實驗室的系統初步檢驗資料，該資訊可於業務簡報時用以證明其有效性。關於逆滲透過程的一般常見問答集，您或許也會覺得有所幫助。

Vocabulary and Phrases

1. transmit [trænsˋmɪt] v. 傳送；傳達
The ship captain transmitted a message using his two-way radio.
船長用他的雙向對講機來傳送訊息。

2. with plenty of time to spare 有充裕的時間
We arrived at the airport with plenty of time to spare.
我們很早就到機場了。

3. reverse osmosis 逆滲透（縮寫為 R / O）
The scientists used reverse osmosis to remove the salt from the sea water.
科學家們使用逆滲透來去除海水中的鹽分。

4. preliminary [prɪˋlɪməˌnɛrɪ] adj. 初步的；預備的
Our preliminary revenue numbers show that we will likely make more of a profit this year.
我們初步的營收數字顯示我們今年可能會賺取更多收益。

5. air cargo 空運貨物
Air cargo arrives in two business days, while regular mail takes more than 15 days.
空運貨物兩個工作天可以送達，而一般貨運要花十五天以上。

6. scrutinize [ˋskrutn̩ˌaɪz] v. 詳細檢查；細看
The jeweler, looking for defects, scrutinized the jade stone.
珠寶商詳細檢查那塊玉石，看看是否有瑕疵。

#2 出貨給客戶的產品彩色盒有瑕疵，於是 Richard 請客戶將毀損的樣品寄回來，同時補寄新的包裝材料給客戶。

TO: Mr. Asahina
FROM: Richard Barlow
DATE: 10/5
SUBJECT: Damaged Display Boxes

Dear Mr. Asahina,

1. We haven't received the samples yet because they were sent as **air cargo**.[5] For security reasons, Taiwan Customs usually **scrutinizes**[6] gas products coming into the country. In the future, we suggest you send samples by an express company such as FedEx or DHL because — although I don't know why — packages sent by this method normally **clear customs**[7] in one day.

2. 90 display boxes were shipped on flight CI106 on 10/4 (MAWB* #297-5218-7284, HAWB* #IEC-901094). Please note that we put an **artificially**[8] low value on the invoice to reduce the amount of tax you pay to customs.

We will arrange to send the remainder of the shipment and advise you of the shipping schedule.

Ciao,
Richard

中譯

#2
收件人：朝比奈先生
寄件人：理查‧巴洛
日期：十月五號
主旨：毀損的展示盒

親愛的朝比奈先生：

1. 我們尚未收到樣品，因為那是用空運寄出的。因為安全考量，台灣海關通常會仔細檢查進口的瓦斯產品。未來我們建議你可以透過像 FedEx 或 DHL 這樣的快遞公司來寄送樣品，因為（雖然我不知道原因）用這種方式寄的包裹通常一天就可以清關。

2. 九十件展示盒已於十月四號經由班機 CI106 寄出（空運主提單編號：297-5218-7284；空運分提單編號：IEC-901094）。請注意為了減低你們付給海關的稅金，我們刻意壓低發票上的貨物價錢。

我們會再安排寄出剩下的貨品並通知你出貨時間表。

再聯絡
理查

7. clear customs 清關

The shipment of soap took a few hours to clear customs.
那批肥皂清關花了幾個小時。

8. artificially [ˌɑrtəˈfɪʃəlɪ] *adv.* 人為地；人工地

The company reported artificially high sales totals.
該公司刻意呈報很高的銷售總額。

Language Corner

Warning about Customs Issues

For security reasons, Taiwan Customs usually scrutinizes gas products coming into the country.

You Might also Say . . .

- For safety concerns, Taiwan Customs tends to examine gas products entering the country very carefully.

▶ Biz Focus

★ MAWB 空運主提單（= Master Air Waybill）；HAWB 空運分提單（= House Air Waybill）

空運主提單為航空公司所簽發之貨運單據，空運分提單則為航空貨運承攬業者根據主提單所簽發之運貨憑據。

Chapter Six

Quality

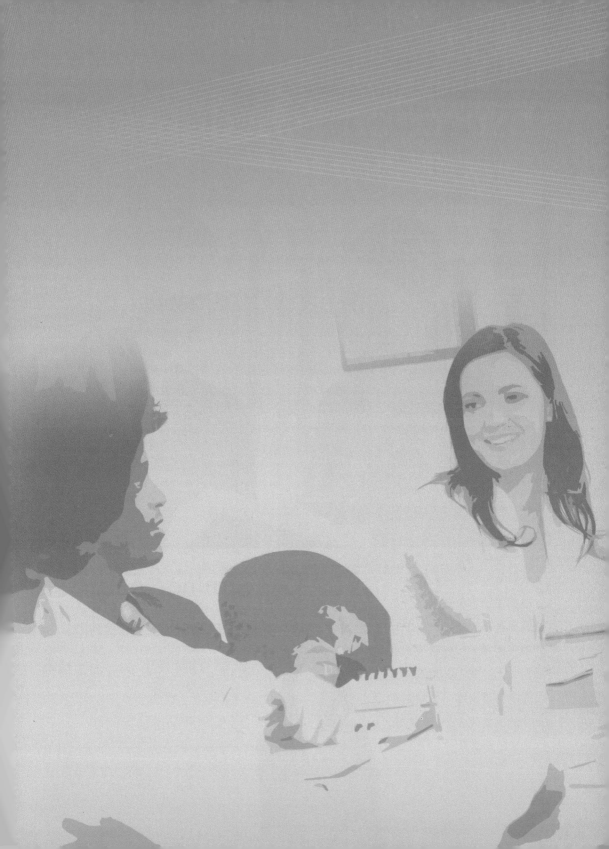

Unit One
Inspection Scheduling
安排檢驗時間

針對新產品或新 **OEM** 工廠的第一次出貨，客戶為慎重其事，通常會要求參觀生產線並親自驗第一批貨，因此賣主必需掌握工廠的生產進度，並聯絡客戶的品管代表於適當時間前來工廠驗貨。

✅ Learning Goals

1. 和供應商敲定檢驗時間

2. 驗貨前和客戶商討品管標準及檢驗計畫

3. 和客戶敲定驗貨時間

Before We Start 實戰商務寫作句型

Topic 1: Getting Information 詢問資訊

* Would you please fill me in on [the details / your progress / the latest developments . . .]?
  可以請你告訴我 [詳情 / 你的進度 / 最新發展……] 嗎？

* I would appreciate your thoughts on the matter / issue / problem.
  你對這件事 / 這個問題的看法如何？

* Any [feedback / suggestions / ideas . . .] is (are) more than welcome / will be greatly appreciated.

* You are more than welcome to provide me with any [feedback / suggestions / ideas . . .].

* I encourage you to share any [feedback / suggestions / ideas . . .] that you may have.
  歡迎提供任何 [回應 / 建議 / 意見……]。

Topic 2: Planning the Schedule 計劃行程

* As I'm sure you can appreciate / understand, . . .
  我確信你可以體諒，……

* What do you think if S. + V. . . .?
  ……，你覺得如何？

* I suggest (that) S. (should) V. . . .
  我建議……

* I'm planning to . . .
  我打算……

A. Scheduling Inspections with Suppliers 和供應商敲定檢驗時間

#1 Maples 台灣分公司總經理 Richard 寫信詢問代工工廠新產品第一次出貨檢驗的時間。

TO: David Chou
FROM: Richard Barlow
DATE: 12/18
SUBJECT: Makeup Mirror Inspection

Dear David,

I trust that the few technical issues **concerning**[1] the mirror will be **settled**[2] soon and that production will **get under way**.[3] I'm planning to come to the Mica Indonesia factory to inspect the first shipment. Would you please **fill** me **in on**[4] MI's holiday schedule for the next few weeks? I would appreciate your thoughts on when you think the best time would be for me to arrive.

Best regards,
Richard

中譯

#1

收件人：周大衛
寄件人：理查‧巴洛
日期：十二月十八號
主旨：化妝鏡檢驗

親愛的大衛：

我相信幾個有關化妝鏡的技術性問題很快就會解決，生產也會開始進行。我打算到 Mica 在印尼的工廠檢驗第一批出貨。可以請你告訴我接下來的幾個星期 MI 工廠什麼時候會放假嗎？你覺得我什麼時候去是最好的呢？

謹上
理查

Vocabulary and Phrases

1. concerning [kənˋsɝnɪŋ] *prep.* 關於
I haven't heard anything concerning the project.
我沒有聽過有關那個計畫的任何消息。

2. settle [ˋsɛtḷ] *v.* 處理；解決
We haven't settled the dispute over the contract.
我們還沒解決關於那份合約的爭端。

3. get under way 開始；進行
Do you know when the conference will get under way?
你知道那個會議什麼時候開始嗎？

4. fill sb in on sth 告知某人消息
They'll fill you in on the details when you arrive.
當你抵達的時候，他們會告訴你細節。

#2 Richard 與工廠協商驗貨時間，希望一次就能驗完兩批貨物。

TO: Benny Yang
FROM: Richard Barlow
DATE: 3/21
SUBJECT: Coming to Indonesia

Dear Mr. Yang,

Thanks for your quick response. As I am sure you can appreciate, I don't want to have someone make two trips to Indonesia to inspect the Company #4S and #2B shipments. **At the present moment,**[5] I'm trying to **figure out**[6] how to inspect both shipments on the same trip.

I don't know if MCUS will start yelling about this idea, but what do you think if our inspector arrives at the MI factory on April 7th and inspects the Company #4S shipment that day? He can then inspect the Company #2B shipment the following two days. Any feedback would be more than welcome.

Sincerely,
Richard

中譯

#2

收件人：楊班尼
寄件人：理查‧巴洛
日期：三月二十一號
主旨：去印尼一事

親愛的楊先生：

謝謝你快速的回覆。我確信你可以體諒，我不想派人跑兩趟印尼，分別去檢查 #4S 和 #2B 公司的貨物。目前我在想該如何一趟就驗完兩批貨。

我不知道 Maples 美國總公司是否會反對，但如果我們的檢查人員四月七號就到 MI 工廠檢查 #4S 公司那批貨，接下來兩天再檢查 #2B 公司的貨物，你覺得如何？歡迎提供任何意見。

祝 商安
理查

5. at the present moment 現在；目前

At the present moment, the company isn't making a profit.
目前那間公司沒有獲利。

6. figure out 想出（解決之道）；理解

I can't figure out why the stock price is dropping.
我想不通為什麼股價一直掉。

Language Corner

How to Confirm the Schedule

I would appreciate your thoughts on when you think the best time would be for me to arrive.

You Might also Say . . .

- Please confirm what you think the best arrival time would be.
- Please let me know when you think the best time would be for me to arrive.
- Please advise me on the best time to get there.

Richard 和新客戶的品管經理約見面，商討出貨前檢驗的細節。

TO: Mr. J. Rusk
FROM: Richard Barlow
DATE: 10/18
SUBJECT: China Inspection Plan

Dear Mr. J. Rusk,

We still do not have a firm production start date for your product, but I think it will occur during weeks 43 or 44. We're awaiting the delivery of some parts. When they arrive, we'll start production, so it may be **on fairly short notice**.[1] I hear that your company wants us to **air-ship**[2] the first order by the end of the month. That means **timing**[3] is important. There will be no pilot run because we have been producing this product since February for other customers. Your version involves only some differently **colored**[4] components. The line capacity* is about 2,000 per day; therefore, the entire production run* should be finished in four or five days.

We'd like to meet with you to review our technology and your inspection plans. Jack Huang from our Taiwan office will be in China for the start of production. I suggest that once a firm schedule is established, he call you to arrange a visit to the factory.

Jack will have an inspection fault list* that has been agreed upon by Company #3P and Maples, plus copies of the versions sheets* for the products. Is there any other information we can provide you to help with your inspection?

Best regards,
Richard Barlow

中譯　收件人：J · 盧斯克先生

寄件人：理查 · 巴洛

日期：十月十八號

主旨：大陸檢驗計畫

親愛的 J · 盧斯克先生：

　　我們還沒有你們公司產品確定的生產日期，但我想會在第四十三或四十四週左右進行。我們正在等一些零件，零件一送到，我們就會開始生產，所以時間會很緊湊。我聽說你們公司要我們在月底前空運第一批貨，那表示時間的安排很重要。從今年二月起，我們已經替其他客戶生產過此項產品，所以將不會有少量的試產。你們公司的版本只有一些零件的顏色不同。生產線的產量大約每天兩千件，所以整個生產作業四到五天就可以完成。

我們想和您碰面，檢視一下我們的技術和你們的檢驗計畫。開始生產時，我們台灣辦公室的黃傑克會到大陸。我建議一旦生產時程確定，他就打電話給您，安排您來參觀工廠。

傑克有 #3P 和 Maples 公司雙方同意的品質瑕疵清單和產品種類型錄表的副本。還有其他資訊是我們可以提供以利您的檢驗嗎？

謹上
理查・巴洛

Vocabulary and Phrases

1. on fairly short notice 在很短的時間內通知

I can fly to Hong Kong on fairly short notice.
我可以在接到通知後很短的時間內就飛去香港。

2. air-ship [ˋɛrˌʃɪp] *v.* 空運

We'd prefer that you air-ship the parts to us.
我們寧願你們用空運將那些零件寄給我們。

3. timing [ˋtaɪmɪŋ] *n.* 時間的選擇、安排

You have the worst timing. I was about to seal the deal when you walked in.
你來的時間很不湊巧。你走進來的時候，我正要敲定那筆生意。

4. colored [ˋkʌlɚd] *adj.* 有顏色的

Our company produces a variety of colored shirts.
我們公司生產各式各樣的彩色襯衫。

Biz Focus

★ **line capacity** 生產線的產量、產能

★ **production run** 生產作業

★ **inspection fault list** 檢驗用的品質瑕疵清單（品質不良表）
　　➡：業界亦普遍使用 faultlist。

★ **versions sheet** （產品的）種類型錄表

Language Corner

Offering Information

Is there any other information we can provide you to help with your inspection?

You Might also Say . . .

▪ What other information could we provide that would help with your inspection?

▪ What else do you need to know that would aid with your inspection?

▪ If you need further information to assist with your inspection, please don't hesitate to contact us.

Unit Two
AQL & Fault List
允收品質標準及品質瑕疵清單

一家公司內各部門的工作執掌差異很大，要讓跨部門的運作得以順利進行，需了解其他部門的工作性質。尤其是代表公司和客戶洽談業務的業務部，更要涉獵一些有關其他部門專業領域的知識，以便回答客戶的問題。

✓ Learning Goals

1. 了解 AQL 的定義，及品管如何根據 AQL 來判定「允收」或「拒收」

2. 了解品管如何根據 fault list 來判別品質是否有瑕疵及缺陷的種類

Before We Start 實戰商務寫作句型

Topic 1: Defining Something 說明定義

- ◆ . . . is basically / typically / most commonly defined as . . .
  ……基本上／通常／最常被定義為……

- ◆ The definition / meaning of . . . is . . .
  ……的定義／意義是……

- ◆ . . . stands for / usually refers to . . .
  ……代表／通常指的是……

- ◆ . . . is viewed as / seen as / thought of as / considered . . .
  ……被視為……

Topic 2: Talking about Methods 談論方法

- ◆ The most common approach to / method of . . . is . . .
  ……最常見的方法為……

- ◆ (Approach) is par for the course / common practice / standard / regular / used a lot / widely used.
  （某方法）很普遍／廣為使用。

- ◆ The inspection plan has / includes / contains / consists of / is made up of . . .
  此檢驗計畫包含……

Richard 向行銷業務部副總裁 Todd 解釋在品管中 AQL 的意涵。

To: Todd Andrews
From: Richard Barlow
Date: 3/13
Subject: Acceptable Quality Level

Dear Todd,

I'm not sure exactly what you're after here, Todd. The AQL, which stands for "Acceptable Quality Level," is basically defined as the worst level of quality that is acceptable to the user. Another definition of AQL is the maximum defective percentage (or maximum number of defects per hundred units) that is considered acceptable as a **process**[1] average.

Once the AQL is established as a requirement, the task is then to try to measure the actual level of quality of a particular lot of goods to see if it satisfies the AQL requirement. The most common approach to doing so is to randomly select a sample of units and inspect them by using a fault list. The Military Standard 105E* inspection plan is **par for the course**[2] in the industry today, but other sampling plans, like the Dodge-Roming plan,* are also used a lot. Every plan provides sampling procedures and tables.*

I hope that **clears** things **up**[3] for you. If you have any more questions or concerns, please don't hesitate to contact me.

Regards,
Richard

中譯　收件人：陶德・安德魯斯
寄件人：理查・巴洛
日期：三月十三號
主旨：允收品質標準

親愛的陶德：

　　陶德，我不確定這究竟是不是你要問的問題，允收品質標準（AQL 代表 Acceptable Quality Level）基本上被定義為使用者所能接受的最差品質。另一個允收品質標準的定義則為所能接受製程平均的最大不良率（或每一百個產品中瑕疵數量的最大值）。

　　一旦建立允收品質標準，接下來的工作就是要試著測量出特定一批貨物的實際品質水準，看是否符合允收品質標準的規定。而其中最常見的方法就是隨機選取樣本，再用品質瑕疵清單來檢驗它們。目前業界使用最廣泛的是美軍標準 105E 抽樣計畫，但其他檢驗計畫如道奇 — 羅敏（抽樣）計畫也很常被使用。每項檢驗計畫均提供抽樣程序及抽樣表。

　　我希望這樣有解決你的疑惑。如果你還有其他問題或疑慮，請儘量和我聯絡。

祝 商安
理查

Vocabulary and Phrases

1. **process** [ˋprɑ͵sɛs] *n.* 過程；進程
   （*v.* 加工；處理）

   They feel that the inspection process takes too much time.
   他們覺得檢驗的過程耗費太多時間。

2. **par for the course** 普遍的

   Their strategy is par for the course in this industry.
   他們的策略在這行很普遍。

3. **clear sth up** 解釋、解決某事

   Those problems will be cleared up soon.
   那些問題很快就會被解決了。

Language Corner

How to Clarify Things

I'm not sure exactly what you're after here.

You Might also Say:

- I don't know if I fully understand your question.
- I'm a little unclear on what exactly your question is.
- I'm not quite sure where your confusion stems from.

Biz Focus

★ **AQL** 允收品質標準（= Acceptable Quality Level）
一般嚴重缺陷的 AQL 定在 0%、主要缺陷定在 1.0%、1.5% 或 2.50%、次要缺陷定在 2.5% 或 4.0%，其中一項達到判退標準產品就拒收，但也有公司是不分種類而使用同樣的 AQL，再將缺陷加總。AQL 的大小視產品的品質要求而定，品質要求越高 AQL 就越小。在新版抽樣計劃中，Acceptable Quality Level 已被改為 Acceptance Quality Limit，中文譯成「允收品質界限」。

★ **Military Standard 105E** 美軍標準 105E 抽樣計畫
二次大戰期間美軍採購的軍品種類繁多，由於先前的檢驗制度無法因應故改用「抽樣檢驗方式」。雖然在 1995 年美國軍方已宣布廢止 MIL-STD-105E，並推薦非官方所制定的 ANSI / ASQC Z1.4-1993，但 MIL-STD-105E 仍廣為民間團體所採行。

★ **Dodge-Roming plan** 道奇 — 羅敏（抽樣）計畫
1924 年美國貝爾電話實驗室之 Walter A. Shewhart 首先將統計應用在品管上，稍後同在貝爾服務之 H. F. Dodge 及 H.G. Roming 於 1944 年提出「道奇 — 羅敏抽樣檢查表」，為現代抽樣計畫的始祖。

★ **sampling procedures and tables** 抽樣程序及抽樣表
如產品「批量（lot or batch size）」是 300 PCS 的話，用「一般檢驗水準 II」，檢驗者可在表 I 查到一個英文字母 H，根據這個代碼可在表 II 得知「樣本數」是 50 PCS。假設買賣雙方所定的 AQL 是 2.5，則「允收數（Ac）」為 3 PCS、「拒收數（Re）」是 4 PCS，即在 50 件中如找到 4 個以上的缺陷則該批貨就驗退拒收。

表 I — MIL-STD-105E 樣本大小代字略表

批量			一般檢驗水準 I	II	III	特殊檢驗水準 S-1	S-2	S-3	S-4
2	至	8	A	A	A	A	A	A	A
9	至	15	A	B	C	A	A	A	A
16	至	25	B	C	D	A	A	B	B
26	至	50	C	D	E	A	B	B	C
51	至	90	C	E	F	B	B	C	C
91	至	150	E	G	H	B	B	C	D
151	至	280	E	G	H	B	C	D	E
281	至	500	F	H	J	B	C	D	E
501	至	1200	G	J	K	C	C	E	F

表 II — 抽樣數量與允收水準 (AQL) （各格內數值為 Ac Re）

樣本數代號	樣本數	0.1	0.15	0.25	0.4	0.65	1.0	1.5	2.5	4.0	6.5	10	15
A	2												
B	3											1	1 2
C	5										1	1 2	2 3
D	8									1	1 2	2 3	3 4
E	13								1	1 2	2 3	3 4	5 6
F	20							1	1 2	2 3	3 4	5 6	7 8
G	32						1	1 2	2 3	3 4	5 6	7 8	10 11
H	50					1	1 2	2 3	3 4	5 6	7 8	10 11	14 15
J	80				1	1 2	2 3	3 4	5 6	7 8	10 11	14 15	21 22
K	125			1	1 2	2 3	3 4	5 6	7 8	10 11	14 15	21 22	
L	200		1	1 2	2 3	3 4	5 6	7 8	10 11	14 15	21 22		
M	315	1	1 2	2 3	3 4	5 6	7 8	10 11	14 15	21 22			
N	500	1 2	2 3	3 4	5 6	7 8	10 11	14 15	21 22				
P	800	2 3	3 4	5 6	7 8	10 11	14 15	21 22					
Q	1250	3 4	5 6	7 8	10 11	14 15	21 22						

B. Fault List 品質瑕疵清單

Richard 發另一封信給 Todd，說明 fault list 的定義。

To: Todd Andrews
From: Richard Barlow
Date: 3/15
Subject: Fault List

Todd,

Further to my letter dated March 13th, a fault list is typically defined as a list of concerns a customer might have about a product.

Specifically speaking,[1] a fault list is a list of performance parameters* that the seller and buyer agree the product should meet. It provides clear, **measurable**[2] **criteria**[3] that an inspector can check to determine whether the product is acceptable or not. Normally, this quality control inspection document has spaces for evaluating the aesthetics of the product. In addition, functional and performance requirements, such as wattage in the case of most electrical appliances, and certainly any and all safety-related **aspects**[4] of the product, are equally important.

I hope my explanation was **satisfactory**.[5] If not, feel free to send me another e-mail and I'll reply as soon as possible.

Cheers,
Richard

中譯　收件人：陶德‧安德魯斯
寄件人：理查‧巴洛
日期：三月十五號
主旨：品質瑕疵清單

陶德：

　　接續我三月十三號發給你的信件內容，品質瑕疵清單通常被定義為一份列有客戶對產品所有可能問題的清單。

　　具體而言，品質瑕疵清單是指買賣雙方都同意產品需達到特定功能設定項目的清單。它提供明確、可測量的標準，讓檢驗人員得以檢查以判定產品是否能被接受。通常這種品管檢驗的文件會有評估產品外觀的欄位。此外，功能與性能規定（例如大多數的電器都要檢驗瓦特數）以及任何與產品安全相關的面向也都很重要。

　　希望我的解釋能讓你滿意。如果還有問題，請再發電子郵件給我，我會儘快回覆。

工作愉快
理查

Vocabulary and Phrases

1. specifically speaking 具體而言；明確地說

Specifically speaking, I was concerned that this product wouldn't sell.

明確地說，我擔心這項產品賣不出去。

2. measurable [ˈmɛʒərəb!] *adj.* 可測量的

We need to find a measurable way to evaluate the performance of our employees.

我們需要找個可測量的方式來評價員工的表現。

3. criteria [kraɪˈtɪrɪə] *n.* 標準；尺度（為 criterion 的複數）

Our products must meet a variety of criteria before they hit the stores.

我們的產品在賣到商店以前必需符合各種標準。

4. aspect [ˈæspɛkt] *n.* 方面；觀點

The most important aspect of our business is customer satisfaction.

我們公司最關切的面向是顧客滿意度。

5. satisfactory [ˌsætɪsˈfæktərɪ] *adj.* 令人滿意的；符合要求的

I'm sure everything was satisfactory.

我確信一切都是令人滿意的。

Biz Focus

★ performance parameter 性能參數（常用複數）

Language Corner

How to Offer Further Help

Feel free to send me another e-mail and I'll reply as soon as possible.

You Might also Say . . .

- Let me know if I can help you in any other way.
- Please don't hesitate to contact me if you need further assistance.
- If you have any more questions, please call or e-mail me at any time.

Unit Three
Fault List Sample
品質瑕疵清單實例說明

本課將提供一份品管檢驗時所使用的品質瑕疵清單範本，根據這份買賣雙方都同意的品管標準，驗貨員可於驗貨時判別產品品質是否有瑕疵及缺陷的種類為何。

✔ Learning Goals

1. 了解實際品管的檢查項目

2. 判別哪些瑕疵會被列為缺陷

3. 辨別嚴重缺陷（與使用安全有關的缺陷）、主要缺陷（功能上的缺陷）及次要缺陷（外觀上的瑕疵，不影響使用效果）

MAPLES / MICA CORPORATIONS
LIGHTED MAKEUP MIRROR FAULT LIST

MAPLES APPROVAL: *Richard Barlow*   DATE: ___*Sept. 30*___
MICA APPROVAL: *Simon Tang*   DATE: ___*Oct. 2*___

FAULT LIST NUMBER: MM-001   REVISION

I. VISUAL INSPECTION	DEFECT CLASSIFICATION
A. Printing / molding defects	Per signed samples
B. Dirty unit	Ditto
C. Missing screw	Major
D. Missing / **illegible**[1] date code	Minor
E. Cracked mirror, but no sharp edges exposed	Major
F. Broken mirror, with edges exposed	Critical
G. Wrong color	Per signed samples
H. **Cosmetic**[2] integrity (scratches, **blemishes**,[3] marks, etc.)	Ditto
I. Missing silkscreening*	Minor
J. **Foreign**[4] material inside unit (**audible**[5] when shaken)	Major

II. FUNCTIONAL CHARACTERISTICS	DEFECT CLASSIFICATION
A. Force required to move the switch from any position to any **adjacent**[6] position shall be 0.2 to 0.9 kgw.	Major
B. Current draw* at 3.0 VDC shall be 1.40 to 1.60 amperes in the "day" setting, 1.30 to 1.40 amperes in the "evening" setting, 0.0 mA in the "off" setting.	This test is only for reference until a specification is agreed upon by Mica and Maples.
C. The switch must turn off when the unit is closed.	Major
D. The battery door must be removable when a force of 0.5 to 1.5 kgw is applied to the latch from a 45-degree angle up and away from the latch.	Major
E. Hinges and pivots shall operate smoothly.	Minor
F. The six assembly screws shall withstand a torque of 1.0 kgw-cm in either direction without turning.	Major

III. PACKAGING	DEFECT CLASSIFICATION
A. Illegible printing / UPC	Major
B. Missing date code on master carton	Minor
C. Damaged or dirty	Per signed samples
D. Missing / wrong instruction book	Major

IV. PHYSICAL INTEGRITY	DEFECT CLASSIFICATION
A. Drop test (units shall be tested opened and closed, but not the same unit. The back cover breaking or **disengaging**[7] shall not be considered a failure.)	
1. The unit shall be functional after one drop from a height of 90 cm onto a hardwood surface.	Major
2. The same unit shall be intact after two additional drops from 90 cm.	Critical

中譯

MAPLES / MICA 公司
附燈化妝鏡品質瑕疵清單

MAPLES 認可：理查·巴洛　日期：九月三十號
MICA 認可：唐賽門　　　日期：十月二號

品質瑕疵清單編號：MM-001　　　　　　　　　　　　　　　　修訂版

I. 外觀檢驗	缺陷類別
A. 印刷 /（塑膠）射出缺陷	根據簽署的樣本
B. 產品有污漬	同上
C. 螺絲鬆脫	主要缺陷
D. 日期代碼不見 / 無法辨識	次要缺陷
E. 鏡子破裂，但銳利邊沒外露	主要缺陷
F. 鏡子破裂，且銳利邊有外露	嚴重缺陷
G. 顏色錯誤	根據簽署的樣本
H. 外觀完整性（刮傷、瑕疵、髒污等）	同上
I. 沒有網印	次要缺陷
J. 產品內含異物（搖晃時可聽見聲音）	主要缺陷

II. 功能特性	缺陷類別
A. 從任何位置搬動開關之力量應介於 0.2 到 0.9 公斤重。	主要缺陷
B. 在 3.0 伏特的直流電壓下，其所消耗電流於「白天設定」應介於 1.40 到 1.60 安培、「夜間設定」應介於 1.30 到 1.40 安培，「關閉」時應為 0.0 毫安培。	在 Mica 與 Maples 雙方確定規格之前，此項測試僅供參考。
C. 鏡子闔上時，開關也需關閉。	主要缺陷
D. 當以 45 度角向閂子施以 0.5 到 1.5 公斤重往上及向外的力量時，放電池的凹槽蓋板應可打開。	主要缺陷
E. 鉸鏈和樞軸需能正常轉動。	次要缺陷
F. 六個組裝用的螺絲需能承受任一方向 1.0 公斤重 — 公分的扭矩而不轉動。	主要缺陷

III. 包裝	缺陷類別
A. 印刷 / 通用產品代碼無法辨識	主要缺陷
B. 外箱沒有日期代碼	次要缺陷
C. 受損或有髒污	根據簽署的樣本
D. 使用手冊遺失 / 有誤	主要缺陷

IV. 物品完整性	缺陷類別
A. 落下測試（分別用不同產品測試鏡子打開和關起來兩種情形。背蓋如有毀損或鬆脫不應視為瑕疵。）	
1. 從九十公分高掉落至硬木表面一次，產品應仍可使用。	主要缺陷
2. 同一件產品從九十公分高再掉下兩次，應仍是沒有損害的。	嚴重缺陷

Vocabulary and Phrases

1. **illegible** [ɪ`lɛdʒəbl] *adj.* 難以辨認的；字跡模糊的

   His handwriting was so terrible that it was nearly illegible.
   他的筆跡很潦草，幾乎無法辨認。

2. **cosmetic** [kɑz`mɛtɪk] *adj.* 表面的；裝飾性的

   Luckily, the damage done by the earthquake was more cosmetic than structural.
   所幸地震造成的損害大多是外觀上，而非結構方面的。

3. **blemish** [`blɛmɪʃ] *n.* 瑕疵；缺點；污點

   Even small blemishes on a retail item can make it tough to sell.
   零售貨品上即使只有些微的瑕疵，就會使它很難賣出去。

4. **foreign** [`fɔrɪn] *adj.* 外來的；無關的

   Even the tiniest foreign object can cause a malfunction.
   即使是一個極小的異物都能造成故障。

5. **audible** [`ɔdəbl] *adj.* 聽得見的

   My coworker speaks so softly that his words are barely audible.
   我同事講話很小聲，幾乎聽不見他說的話。

6. **adjacent** [ə`dʒesṇt] *adj.* 毗鄰的；鄰近的

   The adjacent building is for sale.
   隔壁大樓要賣。

7. **disengage** [ˌdɪsɪn`gedʒ] *v.* （使）鬆開；（使）脫離

   The parts are fragile and might disengage if not handled with care.
   那些零件很脆弱，如果沒有小心拿有可能會鬆脫。

   📓：ditto 指「同上；同前」，用來取代重複的部分。

Biz Focus

★ **silkscreening** 網印（以絲幕上模型複製圖案之方法）

★ **current draw** 消耗電流；電流需求

Notes

Unit Four
Incoming Quality Control
進料（貨）品管

產品是由數十甚至數百種零件組裝而成的，因此任何採購的物料或半成品，一進入倉庫就必需做品管檢驗，以免影響後續加工所製造出來的成品品質。

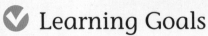

✔ Learning Goals

1. 進料檢驗如發現品質有異時：
   - ▶ 應通知供應商改善，並加強出貨品管
   - ▶ 怕缺料而無法退貨時，需用人工 **100%** 篩選

2. 客戶第一次採購時，需派員協助做進料檢驗，以免因客戶不當檢驗，誤認產品品質不合格

Before We Start 實戰商務寫作句型

Topic 1: Agreeing and Disagreeing 贊成與反對

* I agree / disagree with you that . . .
  我贊成／不同意你……

* We can't come to a consensus on . . .

* We're not on the same page when it comes to . . .

* We don't see eye to eye on / about . . .
  我們在……方面的意見相左。

* I respectfully disagree with your view on . . .
  恕我無法同意你對……的看法。

Topic 2: Apologizing 致歉

* I'm sorry (that) . . .
  我對……感到抱歉。

* I hope you didn't take offense / feel offended.
  我希望你沒有被冒犯到。

* Please forgive / excuse me if S. + V. . . .
  如果……請原諒我。

* Please accept my sincere / heartfelt / genuine apology for . . .
  請接受我對……的誠摯道歉。

Topic 3: Providing Solutions 提供解決方案

* Our corrective action plan for / solution to / approach to this kind of problem is . . .
  我們對這類問題的解決方案是……

* I think the key to the matter / problem / issue (at hand) is . . .
  我想（目前）這個問題的關鍵是……

* This way, we'll be sure / make sure / be certain / know for sure / see to it that . . .
  這樣我們才能確保……

A. Inspection Plan 檢驗計畫

#1 Richard 寫信給在印尼的工廠，質疑為何要取那麼多的樣本來做壽命實驗。

To: David Chou
From: Richard Barlow
Date: 1/20
Subject: Incoming Inspection

Dear Mr. Chou,

I agree with you and James Boston that using the **tightest**[1] AQL for the incoming inspection that the **vendor**[2] will agree to is the best approach. I notice that you are suggesting an S-4 sampling plan* for the bulbs. Do you really plan to **life-test**[3] this many bulbs as part of your incoming inspection?

Best wishes,
Richard

中譯

#1

收件人：周大衛
寄件人：理查·巴洛
日期：一月二十號
主旨：進貨檢驗

親愛的周先生：

我贊成你和詹姆士·波斯頓的想法，採用供應商所同意最嚴苛的品質允收標準來作進料檢驗是最好的做法。我注意到你建議用 S-4 抽樣計畫來測試那些燈泡，你真的打算用這麼多的燈泡來進行壽命實驗，當作你們進貨檢驗的一部分嗎？

謹上
理查

Vocabulary and Phrases

1. **tight** [taɪt] *adj.* 嚴厲的；嚴格的

   Our CEO has some very tight rules about business attire.
   我們的執行長對上班服裝有一些很嚴格的規定。

2. **vendor** [ˋvɛndɚ] *n.* 供應商；賣主

   We plan to buy 300 units from that vendor.
   我們打算和那個供應商買三百件產品。

3. **life-test** [ˋlaɪf͵tɛst] *v.* 測試……的使用壽命

   We need to life-test the samples to determine their durability.
   我們需要測試這些樣本的使用壽命以判定其耐用性。

4. **take offense** 覺得被冒犯（英式拼法為 offence）

   Don't take offense to my comment about your shirt.
   切勿因我對你襯衫的評論而感到被冒犯。

5. **imply** [ɪmˋplaɪ] *v.* 暗示

   He implied that he didn't trust the young assistant.
   他暗示他不信任那位年輕的助理。

6. **internal affair** 內部事務

   Clients should not get involved in our company's internal affairs.
   客戶不應介入我們公司的內部事務。

#2 Richard 在發了前一封信之後,覺得不應介入工廠內部的檢驗工作,因此又再度發信以作澄清。

To: David Chou
From: Richard Barlow
Date: 1/21
Subject: Follow-up on Incoming Inspection

Dear Mr. Chou,

I am sorry that my last e-mail was so poorly worded. I hope you didn't **take offense**,[4] as I didn't mean to **imply**[5] that I agreed or disagreed with your proposal on the S-4 plan. I was merely confirming my understanding of the proposal. I think that MI's incoming inspection plan is your **internal affair**.[6]

Cheers,
Richard

中譯

#2

收件人:周大衛
寄件人:理查‧巴洛
日期:一月二十一號
主旨:有關進貨檢驗的後續

親愛的周先生:

我對我上一封電子郵件寫得非常詞不達意感到抱歉,我希望你沒有被冒犯到,因為我無意暗示我同意或反對你對 S-4 抽樣計畫的提議,我只是在確認我對該建議的了解。我認為 MI 工廠的進料檢驗計畫是你們內部的事務。

收信愉快
理查

Biz Focus

★ S-4 sampling plan S-4 抽樣計畫

Inspection Level(檢驗水準)是用來決定批量和樣本大小的關係,又區分為 General Inspection Level(一般檢驗水準)及 Special Inspection Level(特殊檢驗水準)。後者通常用在可忍受較高抽樣風險而所需樣本數較小的檢驗中,其樣本數會隨著所選用的 S-1、S-2、S-3 或 S-4 而跟著增加。

Language Corner

Showing Respect for Business Partners

I think that MI's incoming inspection plan is your internal affair.

You Might also Say:

- I know that your company's incoming inspection plan is its own private matter.
- I appreciate that you should have full control of the incoming inspection plan.
- I acknowledge that MI should have the final say about the incoming inspection plan.

B. Course of Action 行動方案

#1 Richard 寫信給 Maples 的代工工廠 Global OEM，表達不應由 IQC 人員來篩選進料的不良品，供應商那邊應先做好出貨品管。

TO: Patty Lo
FROM: Richard Barlow
DATE: 9/4
SUBJECT: Defective Parts

Dear Patty,

When I was in your factory, I saw the IQC people **sorting**[1] parts because apparently a few being **yielded**[2] were too short to use. What is your corrective action plan for this kind of problem? Shouldn't the vendor be performing outgoing QC* on these parts?

Whenever you get a chance, it'd be great if you could let me know.

Thanks in advance,
Richard

中譯

#1

收件人：羅佩蒂
寄件人：理查・巴洛
日期：九月四號
主旨：瑕疵零件

親愛的佩蒂：

我在你們工廠時，看到進料檢驗人員在篩選零件，很明顯地一些產出的零件太短而無法使用，妳對這類問題的解決方案是什麼？供應商不用對這些零件做出貨品管嗎？

妳方便的時候，請讓我知道狀況。

先謝了
理查

Vocabulary and Phrases

1. sort [sɔrt] v. 挑選；把……分類

The operators are sorting units that were rejected by our customers.
作業員正在篩選那些被我們客戶退回的產品。

2. yield [jild] v. 出產；產生

You need a good business plan in order to yield a profit.
你需要有好的營運計畫才能獲利。

3. maintain [men'ten] v. 堅稱；主張

Stan Dixson maintained that he was fired for no reason.
史丹・迪克森堅稱他平白無故被開除。

4. in any event 無論如何

In any event, the company will be prepared.
無論如何，公司會做好準備。

5. preferably [`prɛfərəblɪ] adv. 最好；儘量

I'd like to book a flight for sometime next week, preferably Wednesday or Thursday.
我想要訂下禮拜的班機，最好是星期三或星期四。

6. allay [ə`le] v. 平息；減輕

Our top priority right now should be to allay the fears of the investors.
我們現在的首要之務是要減輕投資者的擔憂。

#2 Richard 建議總公司派 QC 人員陪同新客戶 Regist-Tone 做到貨檢驗。

○ ○ ○

TO: Tony Hughes
FROM: Richard Barlow
DATE: 9/4
SUBJECT: Incoming Audit Inspection

Dear Tony,

    Intertime and Bodenham Start both do some sort of incoming audit inspection, while Capertone's has always **maintained**[3] that they do (although I doubt it). **In any event**,[4] I don't think Regist-Tone's request is unusual. I think the key to the matter is how they test and how strictly they interpret the test results. I suggest that a Maples technical representative (**preferably**[5] a QC guy) be present when the first shipment of GasPack® arrives in England, and that he walk the Regist-Tone QC people through the inspection. This way, we'll be sure they're inspecting correctly, and we can immediately **allay**[6] any concerns or questions they'll **undoubtedly**[7] have.

Sincerely,
Richard

中譯

#2

收件人：東尼・休斯
寄件人：理查・巴洛
日期：九月四號
主旨：進料檢驗

親愛的東尼：

    Intertime 和 Bodenham Start 公司都有做某種進料檢驗，Capertone's 公司向來也聲稱他們有這樣做（雖然我有點懷疑）。無論如何我認為 Regist-Tone 公司提出的要求很正常。我想問題的關鍵是他們如何做檢驗，以及用多嚴苛的標準來詮釋檢驗結果。我建議 Maples 的技術代表（最好是品管人員），在第一批 GasPack® 的產品送到英國時可以在場，並和 Regist-Tone 公司的品管人員一起完成檢驗工作，這樣我們才能確保他們正確地檢驗，而我們也能馬上平息他們必定會有的疑慮或問題。

祝 商安
理查

---

7. **undoubtedly** [ʌnˋdautɪdlɪ] *adv.* 毫無疑問地

We undoubtedly have the best inspection team.
毫無疑問地我們擁有最優秀的檢驗團隊。

▶ Biz Focus

★ outgoing QC 出貨品管

Language Corner

How to Ease One's Concerns

We can immediately allay any concerns or questions they'll undoubtedly have.

You Might also Say:

- We can instantly quiet any anxieties they will most certainly have.
- We'll do our best to ease any apprehension they'll be sure to have.
- We'll be able to alleviate all of the worries they'll unquestionably have.

Unit Five
Outgoing Inspection
出貨檢驗

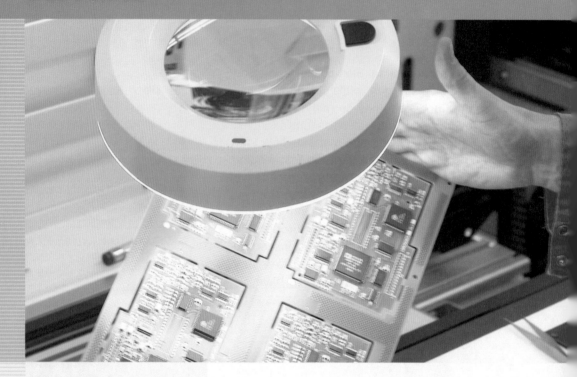

產品必需通過品管檢驗才可以出貨，出貨檢驗可以找出品質的缺失。在允收範圍內可以出貨，否則需驗退、重工。但有時因檢驗標準不明確，或因雙方對不良品的認知有差異，而引發買主和供應商彼此緊張的關係。

✔ Learning Goals

1. 詳述驗退的原因

2. 退貨的不良品經重工後，需通過品管檢驗才可以再出貨，但有時買主選擇相信供應商或非出貨不可，則不重驗

3. 驗貨員不合理的要求，會造成供應商莫大困擾，需透過雙方高層溝通來化解

Before We Start 實戰商務寫作句型

Topic: Giving Reasons 說明理由

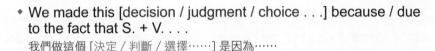

* First / First of all, . . .
  首先，……

* Secondly / In addition / Besides / Furthermore, . . .
  其次 / 此外，……

* We made this [decision / judgment / choice . . .] because / due to the fact that S. + V. . . .
  我們做這個 [決定 / 判斷 / 選擇……] 是因為……

* We [deemed / determined / concluded . . .] it isn't worth + V-ing / it's pointless to V. / there is no point + V-ing / it's a waste of time (money / effort) to V. . . .
  我們 [認為 / 判定 / 推斷……] ……是不值得 / 沒有意義 / 浪費時間（金錢 / 努力）的。

* S. + V. . . .; therefore / thus / as a result / consequently / accordingly, S. + V. . . .
  ……，因此……

* Considering / Since / As S. + V. . . ., I think we're in a position to V. / we ought to be able to V. / we could likely + V. . . .
  由於……，我想我們可以……

Richard 向總公司報告 Maples Taiwan 派驗貨員 Shirley 至印尼做出貨檢驗所發現的問題。

TO: Tony
Cc: James
FROM: Richard
DATE: 9/4
SUBJECT: Makeup Mirror Problems

Dear Tony,

We rejected the Company #2B mirror shipment today for several reasons. First of all, the wrong company name on the units is a clear enough problem. In addition, the problem of noisy hinges is not so **clear-cut**,[1] as it is not specified on the fault list. Now MI wants to put oil on the hinges to quiet them. Shirley is bringing samples back tonight for our review. She thinks that MI already oiled the Company #2B shipment, despite our request that the factory not do anything until we evaluate the samples. **FYI**,[2] Shirley reports that the MI staff has been **uncooperative**[3] and **argumentative**[4] during this trip, and has approached this mirror problem with a very **cavalier**[5] attitude.

Take care,
Richard

中譯 收件人：東尼
副本抄送：詹姆士
寄件人：理查
日期：九月四號
主旨：化妝鏡問題

親愛的東尼：

因幾點原因，我們今天驗退 #2B 公司的化妝鏡。首先，貨品上誤值公司名稱是個很明顯的問題（就足以構成判退）。此外，鉸鏈有聲音的部分是較難界定的問題，因為瑕疵清單上並沒有註明這一項。現在 MI 想要在鉸鏈上塗油以解決噪音的問題。雪莉今晚會把樣品帶回來讓我們複驗。儘管我們已經要求 MI 工廠等我們評估樣本後再採取行動，但她認為 MI 已經在 #2B 公司的貨品上塗了油。也讓你知道一下，雪莉報告說此行 MI 人員很不配合、又愛爭論，處理化妝鏡問題的態度也很傲慢。

保重
理查

► Vocabulary and Phrases

1. clear-cut [`klɪr`kʌt] *adj.* 明確的;清楚的

We have a lot of options and there's no clear-cut way to do this.

我們有很多選擇,這件事並沒有明確的做法。

2. FYI 供你參考(為 for your information 的縮寫,是電子郵件常見用語)

FYI, the shipment is going to be late, so adjust your schedule accordingly.

供你參考,那批貨會晚到,所以你的行程表也要跟著調整。

3. uncooperative [ˌʌnko`ɑpəˌretɪv] *adj.* 不合作的

The customer asked for a refund, but the sales representative was uncooperative.

顧客要求退款,但那名業務員不配合。

4. argumentative [ˌɑrgjə`mɛntətɪv] *adj.* 好爭論的

It's stressful to have a conversation with him because he's so argumentative.

和他對話很有壓力,因為他很喜歡爭論。

5. cavalier [ˌkævə`lɪr] *adj.* 滿不在乎的;傲慢的

I wish you wouldn't be so cavalier about this important matter.

我但願你對這件重要事情的態度可以不要這麼滿不在乎。

Language Corner

1. Explaining a Complex Problem

The problem of noisy hinges is not so clear-cut, as it is not specified on the fault list.

You Might also Say . . .

- The issue of noisy hinges is a little complicated, as it is not included on the fault list.
- The matter of noisy hinges is somewhat unclear because it is not mentioned on the fault list.
- The case of noisy hinges is not easily identified considering it is not detailed on the fault list.

2. Talking about Poor Behavior

The MI staff has been uncooperative and argumentative during this trip.

You Might also Say . . .

- The MI staff has not been cooperating and has argued with us for the whole trip.
- The MI staff has refused to cooperate and has gotten into constant disagreements with us on this trip.

B. Reinspection 複檢

AECO US 下單給 Well Sing 這家工廠，並委託 Richard 的公司 FETS 做出貨檢驗。此封信中 Richard 將向 AECO US 說明不良品重工後不需複檢的理由。

TO: Jim
FROM: Richard
DATE: 10/8
SUBJECT: Well Sing Shipment

Hey Jim,

Well Sing is reworking the rejected shipment for tinning and outer jacket dimensions, and says it will be finished on 10/10.

10/10 is a national holiday in Taiwan. Most companies and government agencies have a three-day weekend. Unfortunately, Taiwan Customs is one of these agencies, so no shipments will leave Taiwan between 10/10 and 10/12. The **bottom line**[1] is these cables will fly out of here on Monday.

I discussed the matter of reinspection with Ken at AECO HK this morning. We agreed not to reinspect anything after the rework. Well Sing is reworking with the understanding that we will reinspect; therefore, we believe that they will in fact rework the shipment with better quality. After Well Sing tells us (probably on Saturday) the reworked shipment is ready for inspection, we'll tell them we're **waiving**[2] it. We made this decision because no matter what the **outcome**[3] of the inspection is, we need to ship. So, we **deemed**[4] that it isn't worth sending an engineer on a six-hour **round-trip**[5] (not to mention the inspection time) during a holiday weekend.

Best regards,
eom

中譯 收件人：吉姆

寄件人：理查

日期：十月八號

主旨：Well Sing 貨品

吉姆，你好：

　　這批退貨 Well Sing 工廠正針對鍍錫以及外層護套的尺寸進行重工的工作，他們說十月十號可以完工。

　　十月十號是台灣的國定假日，大部分的公司與政府機關週末都連休三天。不幸地，台灣海關也是休假的政府機構之一，所以在十月十號到十月十二號期間，所有貨運都無法出關。最重要的是這些電纜線星期一將得空運出去。

我今天早上和 AECO 香港分公司的肯討論複檢一事，我們同意重工後將不再重新驗貨。Well Sing 在認為我們會複驗的情況下進行重工，所以我們相信事實上他們重新整修貨品的品質會比較好。等到 Well Sing 通知我們（大概是星期六）重工的貨品可以驗貨時，我們就會告訴他們不需重新驗貨了。我們會做此決定是因為無論驗貨的結果為何，我們都必需出貨。所以我們認為不值得在週末連假派一個工程師花六小時來回（更別提檢驗的時間）。

謹上

完畢

Vocabulary and Phrases

1. bottom line 最重要的事

The bottom line is we've got to get this thing done by the end of the week.

最重要的是我們必需在本週之前完成這個工作。

2. waive [wev] *v.* 放棄（權利、要求等）；撤回

If you can commit to an order of 50 or more pieces, we'll waive the shipping fee.

如果你可以保證訂五十件或更多，我們就不收運費。

3. outcome [`aʊtˌkʌm] *n.* 結果

They have to wait for the outcome of the first test before moving forward.

在有下一步動作前，他們要先等第一次測試的結果。

4. deem [dim] *v.* 認為；視作

We deemed it unnecessary to hold a meeting without all of the required information.

在沒有備齊所有需要資訊的情況下，我們認為不必要開會。

5. round-trip [`raʊndˌtrɪp] *n.* 來回旅程
（*adj.* 來回的；雙程的）

A round-trip to the capital city takes four hours by bus.

搭公車往返首都要花四小時。

Language Corner

How to Place Emphasis

The bottom line is these cables will fly out of here on Monday.

You Might also Say . . .

- Above all else, these cables need to be shipped out on Monday.
- No matter how it happens, we must deliver these cables on Monday.
- The most important thing is that the cables must be sent out on Monday.

C. Outgoing Inspection Procedure 出貨檢驗程序

由於 #3C 香港分公司所要求的出貨檢驗程序不合理，Richard 發函向總公司報告。

TO: Tony Hughes
FROM: Richard Barlow
DATE: 7/26
SUBJECT: Company #3C Inspection

Dear Tony,

I met with the Company #3C people in HK and their reps in China. We**'re caught up in**[1] a political battle between #3C Taiwan and #3C HK, and we need some intervention from #3C US.

The #3C HK people are **insisting on**[2] an inspection procedure that includes the following steps:

A. A first inspection of each lot of unpackaged products daily by an "inspector" at a Level II sampling.*

B. Once those same lots are packaged, a second inspection by a "supervisor" every three or four days at an S-4 sampling. That means they'll open 32 clamshells for every 2,000 packaged pieces and retest the units.

C. If the S-4 sampling fails, they'll retest the packaged lot at the Level II sampling, — all at no risk or cost to Company #3C.

D. Drop-testing three units from each lot.

E. Life-testing one unit per 10,000 units shipped, and giving the samples and test reports to Company #3C at the end of the test.

F. They want a BOM.*

As I mentioned, #3C HK and #3C Taiwan are **feuding**.[3] #3C Taiwan even complained to its US HQ that #3C HK was just **antagonizing**[4] them and wasting money. Considering each of their demands will cost us money, I think we are in a position to talk to the **higher-ups**[5] at #3C US and tell them that if we have to **abide by**[6] these demands, the unit price must be adjusted.

Regards,
eom

中譯 收件人：東尼‧休斯
寄件人：理查‧巴洛
日期：七月二十六號
主旨：#3C 公司的檢驗

親愛的東尼：

　　我和 #3C 公司的人在香港碰面，也在大陸見到他們的代表。我們被捲入 #3C 台灣與香港分公司之間的紛爭，我們需要 #3C 美國總公司介入協調。

#3C 香港分公司堅持檢驗程序需包含下列步驟：

A. 根據二級抽樣檢驗計畫，每天會有「檢查人員」替每一批未包裝的貨品做第一次檢查。

B. 同一批貨包裝完後，每隔三到四天由「品管主管」依特殊檢驗水準 S-4 級抽樣計畫再做第二次檢查。那意味著每兩千件包裝完的貨品中，他們就會剪開三十二件的雙泡殼，並重驗那些產品。

C. 如果沒通過特殊檢驗水準 S-4 級抽樣計畫，他們會以二級抽樣檢驗計畫再檢查一次，而 #3C 公司無需負擔任何風險或費用。

D. 每一批貨要抽三件來做落下測試。

E. 每一萬件出貨的產品中要抽一件做使用壽命測試，測試結束後需提供 #3C 公司樣本與檢驗報告。

F. 他們要一份物料表。

　　如我先前提到的，#3C 香港與台灣分公司積怨已深。#3C 台灣分公司甚至向美國總部抱怨香港分公司只是在跟他們作對，也很浪費錢。由於他們的每項要求都會耗費成本，我想我們可以向 #3C 美國總公司的高層報告，表達如果我們必需遵照這些要求，貨品的單價就得調整。

祝 商安
完畢

Vocabulary and Phrases

1. be caught up in sth 被捲入某事；被某事牽絆

Keep our deadline in mind and don't be caught up in minor details.
記住我們的截止期限，別被枝微末節給牽絆住了。

2. insist on 堅持

We didn't want to give her a full refund, but she insisted on getting one.
我們不想給她全額退款，但她很堅持。

3. feud [fjud] *v.* 長期爭鬥；爭吵

If you keep feuding with Greg, it might start to distract you from your work.
如果你一直和葛瑞格爭吵下去，可能會開始讓你工作分心。

4. antagonize [æn`tægə‚naɪz] *v.* 敵視；反對

Liz's little brother antagonized her all day.
莉茲的弟弟整天都在跟她作對。

5. higher-up [`haɪə‚ʌp] *n.* 上級；高層

The higher-ups will decide whether Lindsay will keep her job.
公司的高層將決定琳賽是否能保住她的工作。

6. abide by 遵照；信守

The entire operation could fail if she doesn't abide by the inspection plan.
如果她不遵照檢驗計畫，整個運作就會失敗。

Biz Focus

★ Level II sampling 二級抽樣檢驗計畫

在 General Inspection Level（一般檢驗水準）中，可區分為 I、II、III，越後面的樣本數越大，判別力也隨之增大。

★ BOM 物料表（為 Bill of Materials 的縮寫）

Language Corner

How to Describe a Requirement

If we have to abide by these demands, the unit price must be adjusted.

You Might also Say . . .

- If we must conform to these conditions, the unit price will have to be changed.
- If we have to meet / fulfill all of the requirements, the unit price will need to be altered.

Unit Six
Defective Returns 不良品退貨

出貨產品必需通過品管檢驗，但檢驗是隨機抽樣，並不保證消費者買到的產品完全沒問題。因此，退貨的機制非常重要，商家要對消費者負責，而供應商或製造商需對買主或客戶負責。一般而言，買賣合約書會載明百分之幾內的退貨需由買主自行吸收，超過部分才由供應商負責。

 Learning Goals

1. 明確定義「品質不良」

2. 針對一般消費性產品，如消費者退貨比例達 2% 以上，需由供應商負責賠償買主

3. 不管退貨是否已達 2% 以上，買主應每年告知供應商不良品的數量

4. 不良品如不退運重工，而要當場銷毀的話，供應商需派員去監督

5. 若某幾批貨出現全面性品質不良，供應商可要求買主於一定時間（如八個月）內退貨，否則一概不負責

Before We Start 實戰商務寫作句型

Topic 1: Negotiating the Contract 協商合約內容

- I think . . . should be [included in / covered in / stipulated in / written into . . .] the contract.
  我認為合約中應 [包含 / 規定 / 註明……]……

- Basically, I approve of / agree with / support / consent to / have no objections to . . .
  基本上我同意……

- I'd like to see / put (throw) in / write a stipulation in the contract that [requires / asks / specifies / details / lists . . .] . . .
  我希望合約上納入 / 加註條文，[要求 / 詳細說明 / 列舉……]……

- We should have the option / choice of . . .
  我們應有……的選擇權。

Topic 2: Clarifying Responsibilities 釐清責任歸屬

- If the customer fails to . . . , we should be absolved from (of) any responsibility / excused from all liability / free of blame / off the hook for . . .
  如果客戶未能……，我們無需對……負責。

- . . . is responsible for / has responsibility for the [verification / resolution / destruction / evaluation / inspection . . .] of defective returns.
  ……有責任要 [確認 / 解決 / 銷毀 / 評估 / 檢查……] 不良品退貨。

A. Defective Return Policy 退貨機制

Richard 寫信給總經理，希望和客戶簽訂的合約書中，能訂定產品品質標準及不良品退貨的機制。

TO: Tony Hughes
FROM: Richard Barlow
DATE: Oct. 5
SUBJECT: Contract Requirements

Dear Tony,

I think that the defect definitions, AQLs for each definition, an inspection agreement (detailing who inspects the product, where, how, etc.), and a defective return policy should be included in the contract.

Basically, I approve of the Company #3C defective return policy (two percent), except there still aren't any **enforcement**[1] controls in it. I'd like to see a **stipulation**[2] in the contract that requires the customer to report to us at least once a year the number of returns they have **in stock**.[3] In addition, we should have the option of **verifying**[4] the defective units at the customer's factory / warehouse whenever we feel it is necessary (whether the rate is over two percent or not). If the customer fails to advise us of their returns in any year, I think this should **absolve** us **from**[5] any responsibility for those units.

An agreed-upon definition of a defective return would be nice to have too so we don't **run into**[6] another Company #3C-type problem of **making up**[7] defects that aren't on the fault list, or judging a used unit by new unit standards.

If you have **objections**[8] to any of my proposed changes, please let me know and we'll find a way to work them out.

Thanks,
Richard

中譯 收件人：東尼・休斯
寄件人：理查・巴洛
日期：十月五號
主旨：合約規定

親愛的東尼：

　　我認為合約中應包含瑕疵的定義、每項定義的允收品質標準、檢驗協議書（詳細指出誰來檢驗、檢驗的地點及方式等），以及不良品退貨的規範。

　　除了缺少強制執行的規定外，基本上我同意 #3C 公司的退貨規範（百分之二）。我希望合約納入「要求客戶至少每年需向我們回報一次有多少退貨庫存」的條文。此外，每當我們認為有必要時，我們應有至客戶的工廠 / 倉庫查驗不良品的選擇權（無論退貨率是否超過百分之二）。若客戶在任何

一年未能告知退貨情形，我認為我們無需對那些退貨負責。

雙方能就不良品退貨的定義達成共識是很好的事，這樣可避免類似 #3C 公司捏造品質瑕疵清單上沒有的缺陷，或以新產品的品質標準來判別使用過的舊產品等問題。

如果你對我提議的修改有異議，請讓我知道，我們可以設法來解決。

謝謝
理查

➤ Vocabulary and Phrases

1. enforcement [ɪnˈfɔrsmənt] *n.* 強制；執行

Quality-control standards are worthless without proper enforcement within the factory.
工廠內若沒有適當的執行，品管標準就毫無用處。

2. stipulation [ˌstɪpjəˈleʃən] *n.* 條文；規定

The stipulation states that the amount must be paid before the item is delivered.
這項條文聲明必需在運送物品前付費。

3. in stock 庫存

I'll check in our warehouse to see how many appliances we have in stock.
我會檢查我們的倉庫看還有多少庫存的設備。

4. verify [ˈvɛrəˌfaɪ] *v.* 證實；查核

I need to check with my boss to verify that guests are allowed in the factory.
我必需向老板確認客人是否獲准進入工廠。

5. absolve sb from sth 免除某人的責任、義務

If the application isn't submitted, the company will be absolved from all liability.
如果沒有提出申請的話，公司將不負任何責任。

6. run into 陷入困境；遭遇麻煩

Our competitor is running into financial difficulties.
我們的競爭對手正遭逢財務困難。

7. make up 捏造

He's making the whole thing up. Don't listen to him.
這整件事都是他捏造的。別聽他的話。

8. objection [əbˈdʒɛkʃən] *n.* 反對；異議

She had an objection to the businessman's opinion.
她反對這位商人的意見。

Language Corner

Protecting One's Rights

We should have the option of verifying the defective units whenever we feel it is necessary.

You Might also Say . . .

- We should have the right to verify the defective units whenever we feel it is needed.
- We should be able to choose whether or not to verify the defective returns whenever we feel it is warranted.
- We should have the freedom to decide whether or not to verify the faulty products whenever we deem it necessary.

B. Defective Return Audit 不良品退貨查核

Richard 希望總公司能派員查核客戶遭消費者退貨的不良品，並監督不良品的銷毀過程。

TO: Tony Hughes
FROM: Richard Barlow
DATE: 8/12
SUBJECT: **On-site**[1] Audit

Dear Tony,

I don't remember if you were with the group when we discussed the destruction of the returns, but Company #3C has agreed to have a Maples rep on-site to **witness**[2] the destruction of all the returns. Given that Company #3C has recently attempted to resell the defective products (which is against the terms of our contract), it might be a good idea to push that appointment along.

I'll start the arrangements with Company #3C as soon as I receive your confirmation to proceed. By the way, I never did learn the outcome of our last audit at Company #3C. Were those **results**[3] settled and agreed upon?

Send me an e-mail at your earliest convenience. I appreciate all your help in this matter.

Take care,
Richard

中譯

收件人：東尼・休斯
寄件人：理查・巴洛
日期：八月十二號
主旨：現場查核

親愛的東尼：

我不記得當我們討論銷毀不良品時，你是否有與會，但 #3C 公司同意讓 Maples 公司的代表在場目睹銷毀全部的瑕疵品。有鑒於最近 #3C 公司試圖要重新販售那些不良品（這違反我們的合約條款），促成這項約定或許是個好主意。

一旦和你確認可以繼續進行後，我就會與 #3C 公司著手安排相關事宜。對了，我都還不知道我們上次在 #3C 公司稽查的結果為何，結果已經確定並得到雙方同意了嗎？

方便的話，請儘早發信回覆。我很感激你對此事的幫忙。

保重
理查

Vocabulary and Phrases

1. on-site [ɑn`saɪt] *adj.* 現場的（亦可當 *adv.*）
We're going to conduct an on-site inspection.
我們將會進行現場檢驗。

2. witness [`wɪtnɪs] *v.* 親眼看見；目睹
I witnessed the theft of the television.
我目睹了那台電視被偷。

3. result [rɪ`zʌlt] *n.* 結果
The results of the survey were very surprising.
調查的結果令人感到十分意外。

4. resolution [ˌrɛzə`luʃən] *n.* 解決；解答
We need to find a resolution for this problem this week.
我們必需在這禮拜找出這個問題的解決辦法。

5. epidemic [ˌɛpə`dɛmɪk] *adj.* 普遍的；流行的
Due to epidemic defects, all of the radios were returned to the manufacturer.
由於有普遍性瑕疵，所有的收音機都退回給製造商。

6. off the hook 解除困難或義務；擺脫困境
Because it wasn't the company's fault, they were off the hook for the refunding of the parts.
因為這不是該公司的錯誤，所以他們無需對那些零件的退款負責。

C. Return of Defective Products 不良品退運

Richard 去函總經理以了解為何未做稽查，客戶就將不良品退回工廠，以及總公司和 OEM 工廠針對客戶退貨的協議書內容。

TO: Tony Hughes
FROM: Richard Barlow
DATE: 8/16
SUBJECT: Company #4P Defective Returns

Tony,

I read something in the faxes from Mica about Company #4P returning a container of defective products to Mica. Do you know what this is all about? The last I heard, Maples QA department is responsible for the verification and **resolution**[4] of defective returns at the customer's site before anything is returned. Please enlighten me on the details of this current situation.

Mica also tells us that a couple of months ago Maples US signed a new agreement with them that says **epidemic**[5] defects must be returned to Mica within eight months, or Mica is **off the hook**.[6] Is this correct? May we have a copy of the agreement?

Thanks for looking into these matters for me.

Ciao,
Richard

中譯

收件人：東尼‧休斯
寄件人：理查‧巴洛
日期：八月十六號
主旨：#4P 公司的不良品退貨

東尼：

我在 Mica 發出的傳真上看到 #4P 公司退了一貨櫃的不良品到 Mica 工廠。你知道這是怎麼一回事嗎？就我最近所知的消息是：在退貨前，Maples 的品保部門有責任要確認與解決客戶那邊的不良品退貨。請讓我知道目前的詳細狀況。

Mica 還告訴我們幾個月前 Maples 美國總公司和他們簽定新合約，上頭註明若某某批貨出現全面性品質瑕疵需在八個月內退回 Mica 工廠，否則 Mica 無需負責。這是正確的嗎？可以給我們一份協議書的副本嗎？

謝謝你幫我調查這些事。

再聯絡
理查

Language Corner

1. How to Explain Your Reasoning
+ Given . . .
+ Taking into account . . .
+ Considering . . .
+ Due to . . .

(註) 上述字詞後面可接名詞或 that 子句，但需注意 due to 後面若要接子句，需說成 Due to the fact that . . .。

For example:
1. Taking into account the extra cost of shipping, the total will be $541.54.
2. They needed to work overtime due to the fact that the shipment had to be out by Tuesday.

2. How to Bring One Up to Date
+ The last I heard, . . .
+ As far as I know, . . .
+ For all I know, . . .
+ To the best of my understanding / knowledge, . . .

For example:
1. The last I heard, they expected the shipment to be on time.
2. As far as I know, the products reached the factory yesterday.

Chapter Seven

Travel / Meetings

Unit One
Business Travel Plan /
Itinerary 出差計畫 / 行程安排

出差拜訪客戶除了能達到面對面溝通的目的，也可藉此建立私誼，以利推展未來業務。出差行程的安排要先考量自己的工作，再向受訪者提出建議拜訪行程，也可直接委由受訪者來作安排。

✔ Learning Goals

1. 和客戶敲定碰面時間及相關行程細節

2. 處理突發狀況或行程衝突

3. 請旅行社針對規劃的行程來作報價

Before We Start 實戰商務寫作句型

Topic 1: Scheduling 安排時間

* We have that entire [morning / afternoon / evening / period . . .] free.
  我們那整個 [早上 / 下午 / 晚上 / 時段……] 都有空。

* We will have some time available / free in the A.M. (morning) / P.M. (afternoon) [if we need it / just in case / if there is something else to discuss . . .].
  [如果需要的話 / 以防萬一 / 假如有其他事要討論……]，我們早上 / 下午有一些空檔時間。

* (Date / Time) is open for / good for / fine by me.
  我（日期 / 時間）有空。

* How does (date / time) work for / sound to you?
  你（日期 / 時間）可以嗎？

* In case something pops up / comes up / happens / arises, you can reach me [at my office / on my cell phone / by (via) e-mail (fax) . . .].
  如果有事發生，你可以 [打到我辦公室 / 打我的手機 / 透過電子郵件（傳真）……] 和我聯繫。

* He'd like to rendezvous / meet up / set up a meeting / get together / meet head-to-head with you.
  他想要和你見個面。

Topic 2: Planning Tours 計劃旅遊

* How much will / does it cost to fly business class / economy class / first class?
  搭商務艙 / 經濟艙 / 頭等艙要多少錢？

* He's interested in scoping out / checking out / visiting / viewing / seeing . . .
  他對參觀……有興趣。

* I'll need [airfare / hotel room / rental car / limousine . . .] quotes.
  我將需要 [飛機票價 / 飯店房間 / 租車 / 大型豪華轎車……] 的報價。

To: Ken Smith
From: Richard Barlow
Date: May 27
Subject: **Hooking Up**[1]

Dear Ken,

I enjoyed meeting with you the other day, but we still didn't get through all of our **agenda**.[2] Thus, I look forward to **picking up**[3] our discussion in Hong Kong.

Jack and I will be in Hong Kong next week, and I hope we can arrange a meeting. We're arriving late morning on June 4; we have that entire afternoon and evening free. Can you **free up some space**[4] for us then?

June 5, 6, and 7 are going to be **hectic**[5] for us, but we will have some time available in the A.M. on the 8th if we need it.

Sincerely,
Richard

To: Richard Barlow
From: Ken Smith
Date: May 27
Subject: Re: Hooking Up

Dear Richard,

Nice to hear that you and Jack will be able to visit us in Hong Kong. Feel free to come by our office anytime. Or maybe we could do dinner. I'll invite my partner, Jim Wood, along as well. You two still haven't been introduced to Jim.

On another note,[6] I was unaware that we still had some business to take care of. June 4th is open for me. Perhaps it would be best if I sent a driver over to your hotel. How does four o'clock work for you?

In case something **pops up**,[7] my office number is 2574-7722. Or you can reach me on my cell phone. You already have the number.

I hope we can **touch base**[8] like this more often in the future.

Regards,
Ken

收件人：肯‧史密斯
寄件人：理查‧巴洛
日期：五月二十七號
主旨：見面

親愛的肯：

我很開心那天能和你見面，但是我們尚未討論完所有的事項。因此我期待我們能在香港繼續我們的討論。

我和傑克下週會在香港，我希望我們可以安排時間見面。我們會在六月四號接近中午的時候抵達；我們那整個下午和晚上都有空。你到時候可以空出一點時間給我們嗎？

六月五號、六號和七號我們都會很忙，不過如果有需要的話，我們八號早上有一些空檔時間。

祝 商安
理查

收件人：理查‧巴洛
寄件人：肯‧史密斯
日期：五月二十七號
主旨：回覆：見面

親愛的理查：

很高興聽到你和傑克將來香港拜訪我們。歡迎隨時過來我們的辦公室。或者也許我們可以一起吃晚餐。我也會邀請我的夥伴吉姆‧伍德一同出席。還沒介紹你們和吉姆認識。

順帶一提，我不知道我們還有事情沒有處理完。我六月四號有空。或許我派一位司機到你們的飯店會是最好的。你們四點可以嗎？

如果有事發生，我辦公室的電話號碼是2574-7722。或者你可以打我的手機。你已經有號碼了。

我希望我們未來可以像這樣常聯絡。

祝 商安
肯

Vocabulary and Phrases

1. **hook up** 碰面
   Let's hook up after work for drinks.
   我們下班後碰個面喝一杯吧。

2. **agenda** [ə`dʒɛndə] *n.* 議程；會議事項
   We have a full agenda today.
   我們今天的議程很滿。

3. **pick up** 再繼續
   I was hoping to pick up where we left off last time.
   我希望我們可以繼續上一次停下來的話題。

4. **free up some space** 空出一點時間
   Mr. Barry promises to free up some space to meet your representative before he leaves for Paris.
   貝利先生答應在他去巴黎之前空出一點時間和你們的代表見面。

5. **hectic** [`hɛktɪk] *adj.* 忙亂的
   Working at a newspaper, you'll have to meet deadlines. It's quite hectic.
   在報社上班你得在截稿時間內完成工作。那是很忙亂的。

6. **on another note** 順帶一提
   John is leaving on the 31st; on another note, tell us if you know a good person to replace him.
   約翰三十一號離開；順帶一提，如果你認識可以替代他的好人選，要告訴我們。

7. **pop up** 突然發生、出現
   Dennis isn't attending the conference call. Something popped up.
   丹尼斯沒有參加電話會議。有事突然發生。

8. **touch base** 聯繫
   I haven't seen you for a long time. We should touch base more often.
   我好久沒見到你了。我們應該更常聯繫。

Language Corner

Checking One's Schedule

Can you free up some space for us then?

You Might also Say . . .

- Please make some time available for us.
- Put some time aside for us.
- Could you please make yourself available?

B. Changing Plans 變更計畫

TO: Jan Meyer
FROM: Alex Ramirez
DATE: September 25th
SUBJECT: Your Arrival

Dear Jan,

Something has suddenly come up. I thought I'd let you know what is happening so you can plan your trip accordingly.

First, it looks as if I'll be heading to the States for about two weeks. I didn't know this yesterday when I talked with you, and it is not settled yet, but it looks very likely. Considering this, you should plan to arrive in early November.

By the way, do you **reckon**[1] there is any need for me to stop in England during my return trip to Taiwan? Are there any technical or QC issues to clear up with Mr. Kelp or Mr. Harris — or any **hitches**[2] in Project BT? **Say the word**[3] and I'll be more than happy to help out.

Sincerely yours,
Alex Ramirez

中譯

收件人：珍・梅耶
寄件人：艾力克斯・拉米瑞茲
日期：九月二十五號
主旨：妳的到訪

親愛的珍：

突然有事發生。我想我要讓妳知道發生什麼事，這樣妳好依此規劃行程。

首先，看起來我似乎將要去美國約兩個禮拜。我昨天和妳談話時還不知道這件事，它還沒確定，但看起來非常有可能。有鑑於此，妳應該計劃十一月初抵達。

對了，妳覺得我回台灣時需要中途停留英國嗎？有什麼技術或品管上的問題需要向凱爾普先生或哈利斯先生釐清的嗎？或是 BT 案有任何的問題嗎？只要妳一句話，我會很樂意幫忙的。

謹上
艾力克斯・拉米瑞茲

Vocabulary and Phrases

1. **reckon** [ˋrɛkən] v. （口語）覺得；猜想

   Do you reckon that he'll talk about the annual report during the meeting?
   你覺得他會在會議上談論年度報告嗎？

2. **hitch** [hɪtʃ] n. 障礙；故障

   Our client has agreed to the terms, but there are a couple of hitches.
   我們的客戶已經同意這些條件了，但還有一些問題。

3. **say the word** 一句話

   If you want us to come to your wedding, just say the word.
   如果你想要我們參加你的婚禮，就說一聲。

4. **conflict** [ˋkɑnflɪkt] n. 抵觸；衝突

   I hope this doesn't cause a conflict in your schedule.
   我希望這不會和你的行程表衝突。

C. Making an Itinerary 規劃行程

To: Karen
From: Scott
Date: September 30th
Subject: Travel Plans

Dear Karen,

I discussed your travel plans with Tony and there may be a **conflict**.[4] Tony will travel to Southeast Asia via Hawaii, departing from Los Angeles on October 22nd. During his **stopover**,[5] he'll attend a **merchandise**[6] trade show and inspect our operations. He'll proceed to Hong Kong on the 28th, where he'd like to rendezvous with you. Here's the rest of his itinerary: Shenzhen on the 29th and Taipei on the 30th.

There is no problem with you coming to the US. Tony suggests you schedule your travels around the Thanksgiving holiday. It's the ideal time, as there's a **lull**[7] before we head into the Christmas season. In any case, be sure you're there when Tony is there and that Tony is here when you are here.

Regards,
Scott

中譯

收件人：凱倫
寄件人：史考特
日期：九月三十號
主旨：旅遊計畫

親愛的凱倫：

我和東尼討論過妳的旅遊計畫，可能會有衝突。東尼將在十月二十二號離開洛杉磯，途經夏威夷最後抵達東南亞。在他中途停留的期間，他將會參加一場商品貿易展並檢視我們的營運。二十八號他會到香港，他想在那裡和妳見面。他後續的行程為：二十九號在深圳、三十號在台北。

妳要來美國是沒有問題的。東尼建議妳將行程安排在感恩節假期期間。那個時間很理想，因為我們在進入耶誕季前活動會比較少。無論如何，請確認妳和東尼會在同一個地方。

祝 商安
史考特

---

stopover [ˋstɑpˏovɚ] *n.* 中途停留
It's a direct flight. There are no stopovers.
這是直飛的班機，中途沒有停留。

merchandise [ˋmɝtʃənˏdaɪz] *n.* 商品；貨物
To celebrate its 15th anniversary, the company will be giving away free merchandise.
為了慶祝十五週年，該公司將贈送免費商品。

lull [lʌl] *n.* 暫時平息；間歇
There was a lull in the storm, but then it started raining hard again.
暴風雨暫時平息，但之後又開始下大雨。

Language Corner

Unexpected Rescheduling
Something has suddenly come up.

You Might also Say . . .

- I've been called away on urgent business.
- We've had a last-minute turn of events.
- This just came out of the blue.

#1

To: Global Travel
From: Maggie Chen
Date: April 23
Subject: Inquiring about Tickets

To Whom It May Concern:

1. My employer wants a round-trip ticket to San Diego.

2. My employer would like to spend a week in Europe. He'd prefer a **tour package**[1] that includes Nice, Florence, and Geneva. He's also interested in **scoping out**[2] secondary cities and is fairly **flexible**[3] on this.

Please tell me about tour packages you have that are close to fitting his needs. I'll also need **airfare**[4] quotes for both business and economy class.

Thank you,
Maggie Chen

中譯

#1

收件人：全球旅遊
寄件人：陳瑪姬
日期：四月二十三號
主旨：詢問機票

敬啟者：

1. 我的老闆要前往聖地牙哥的來回機票。

2. 我的老闆想要在歐洲待一個星期。他偏好包含尼斯、佛羅倫斯和日內瓦在內的套裝行程。他對參觀次要的城市也有興趣，且對此安排的彈性很大。

請告訴我所有接近他需求的套裝行程。我還需要商務艙和經濟艙飛機票的報價。

謝謝你
陳瑪姬

Vocabulary and Phrases

1. **tour package** 套裝（旅遊）行程

Do the tickets to the opera come as part of this tour package?
歌劇的票有包含在這個套裝行程嗎？

2. **scope out** 參觀

Let's go for a walk and scope out the neighborhood.
我們去附近走走參觀一下吧。

3. **flexible** [ˋflɛksəbḷ] *adj.* 有彈性的；可變通的

In order to get this deal done, you'll have to be more flexible.
為了完成這個交易，你需要更有彈性。

4. **airfare** [ˋɛrfɛr] *n.* 飛機票價

Whenever oil prices go up, airfares become m expensive.
每當油價上漲時，飛機票價都會變得更貴。

5. **take sb up on sb's offer** 接受某人的提議、出價

Maggie says she'll take you up on your offer f lunch next week.
瑪姬說她接受你下禮拜的午餐邀請。

6. **treat** [trit] *n.* 樂事；樂趣

You ordered pizza? What a treat.
你訂了比薩？真棒。

#2

TO: John Short
FROM: Tiffany Liao
DATE: 1/31
SUBJECT: We'll Be in London Soon

Dear Mr. Short,

Thank you for your letter of January 29. Your proposed itinerary sounds fine. We will arrive at your factory mid-morning on February 17. It is probably best if we plan to check into a hotel near your location for that night, so if I can still **take you up on your offer**,[5] would you please book a room for us?

My husband, Mike, and I thank you very much for arranging a tour of Windsor Castle. It sounds like it'll be a **treat**,[6] especially with your secretary leading the way. However, I must say that if she finds she is needed in the office at that time, Mike and I will be fine on our own. We're both quite **resourceful**.[7]

I'll be in our San Jose office for the next few days. I can be reached at (408) 535-7638 or by e-mail. I'm looking forward to meeting you.

Best regards,
Tiffany Liao

#2

收件人：約翰・蕭特
寄件人：廖蒂芬妮
日期：一月三十一號
主旨：我們很快會到倫敦

親愛的蕭特先生：

謝謝您於一月二十九號的來信。您建議的行程聽起來不錯。我們將在二月十七號上午十點左右到達您的工廠。假如那天晚上我們計畫入住你們附近的飯店可能會是最好的。如果我還可以接受您的提議，可以請您幫我們訂房嗎？

我和我的先生麥克非常感謝您為我們安排溫莎堡的行程。這行程聽起來非常好玩，特別是有您的秘書當嚮導。不過，如果她到時候發現她必需待在公司，我和麥克自己去就可以了。我們都非常善於應變。

接下來的幾天我會在我們聖荷西的辦公室。您可以打電話到 (408) 535-7638 或寄電子郵件和我聯絡。我期待與您見面。

謹上
廖蒂芬妮

resourceful [rɪˋsɔrsfəl] *adj.* 善於應變的；足智多謀的

I don't speak Polish, but don't worry about it. I'm a pretty resourceful guy.
我不會說波蘭語，但是別擔心。我是一個很會應變的人。

Language Corner

Inquiring about a Service

Please tell me about tour packages you have that are close to fitting his needs.

You Might also Say . . .

- Please tell me if you have anything that fits that description.
- Do you have something along those lines?

Unit Two
Factory Visit 參觀工廠

客戶參訪工廠的觀感對雙方日後的合作機會有關鍵性的影響，因此規劃參訪行程需十分謹慎，務必確保相關部門有充足的時間來作準備。

 Learning Goals

1. 提前通知工廠客戶來訪的日期、目的及需準備的事項

2. 工廠對買主帶來的客人要緊守分際，不可越俎代庖

3. 客人於參訪後應寫信感謝對方的招待，並分享參訪心得

Before We Start 實戰商務寫作句型

Topic: Discussing a Business Visit 討論商務參訪

* Please be advised / note / remember that (someone) will be coming / visiting / arriving on (date) for a tour of our factory / plant / works.
  請注意／記得（某人）將在（日期）造訪我們的工廠。

* I will notify you about / advise you of / inform you of / let you know (someone's) [exact / finalized / revised . . .] itinerary / schedule as soon as I get it.
  我一收到（某人）[確切／最後確定／修改過的……] 旅行計畫／行程就會通知你。

* The purpose / goal / intention of this [trip / visit / call / appointment . . .] is to . . .
  此次 [行程／參訪／約會……] 的目的為……

* Please remember / keep (bear) in mind / don't forget that (someone) needs to V. . . .
  請記得（某人）必需……

* I believe the visit [went well / was a success / bore fruit . . .], with the exception of / except (for) + N.
  除了……，我相信這次的參訪 [很順利／成功……]。

* Here are my observations / comments / opinions / remarks . . .
  這是我的看法／意見……

* I consider this trip [a waste of my time / a fruitless effort . . .].
  我認為此行 [是在浪費我的時間／毫無收穫……]。

A. Welcoming Customers 歡迎客戶

#1 Kenneth 通知工廠客戶即將來訪的消息。

TO: Tommy Boyle
FROM: Kenneth Davis
DATE: March 25th
SUBJECT: Stuffer's Cookies

Dear Tommy,

Please be advised that Jack Shaw, Paul Alexander (Stuffer's technical manager), and Edwin Meredith (a Stuffer's customer in England) will be coming to Taiwan on April 6 for a tour of our factory. I will notify you about their exact itinerary as soon as I get it.

In addition to showing Edwin the production line, the purpose of this trip is to **iron out**[1] some quality issues they have. Please remember that your QC people need to be well prepared for the upcoming meeting with them and able to respond to their quality concerns.

Last but not least, we need to put up some placards to welcome our guests. I definitely want one in the lobby and another in the **boardroom**.[2] Furthermore, make sure that we have enough company info **kits**[3] **on hand**.[4]

Thank you for your assistance.

Sincerely,
Kenneth

中譯

#1
收件人：湯米‧波義爾
寄件人：肯尼斯‧戴維斯
日期：三月二十五號
主旨：Stuffer's Cookies

親愛的湯米：

請注意蕭傑克、保羅‧亞歷山大（Stuffer's 的技術經理）和艾德恩‧梅瑞迪斯（Stuffer's 在英國的客戶）將會在四月六號來台灣參觀我們的工廠。我一收到他們確切的行程表就會通知你們。

除了帶艾德恩參觀生產線之外，此行的目的是要解決他們一些有關品質的問題。請記得你們的品管人員必需對即將到來的會面有萬全的準備，並能夠回應他們所關心的品質問題。

最後但並非最不重要的一點，我們需要張貼一些海報來歡迎我們的貴賓。我當然想要放一張在大廳，再放一張在會議室。再者，要確認我們手邊有足夠的公司介紹資料袋。

謝謝你們的協助。

祝 商安
肯尼斯

Vocabulary and Phrases

1. iron out 解決；消除

I need to iron out some problems with my presentation before I give it next week.
我必需在下星期簡報前解決一些問題。

2. boardroom [`bɔrd͵rum] *n.* 會議室

Today's meeting will be held in the boardroom.
今天的會議將在會議室舉行。

3. kit [kɪt] *n.* 一套（說明書等）資料

Each kit includes a company overview, notepad, and pen.
每套資料裡都有一份公司簡介、一本筆記本和一支筆。

4. on hand 在手邊；在近處

We have 25 staff members on hand to assist you.
我們現在有二十五位工作人員可以協助你。

5. pass sth with flying colors 某事非常成功

Rudy passed the interview with flying colors.
盧迪面試非常成功。

6. kudos [`ku͵dɑs] *n.* 讚揚；名聲

Kudos to you and your department for all of the great work you've done recently.
所有你和你部門最近出色的工作表現都值得讚賞。

#2 參訪行程圓滿落幕，Kenneth 發信感謝工廠的配合。

TO: Tommy Boyle
FROM: Kenneth Davis
DATE: 4/7
SUBJECT: **Passing with Flying Colors**[5]

Dear Tommy,

Kudos[6] to you and your staff for making **Messrs.**[7] Shaw, Alexander, and Meredith's factory visit a success. They enjoyed meeting with you and learned a lot from what they saw in our factory. Mr. Alexander has indicated that his report to his company will be a very **favorable**[8] one. Your staff's preparation for this visit was obvious. So go ahead and give yourselves a pat on the back.

Well done!

Kenneth

中譯

#2

收件人：湯米・波義爾
寄件人：肯尼斯・戴維斯
日期：四月七號
主旨：非常成功

親愛的湯米：

你和你的同仁們讓蕭先生、亞歷山大先生和梅瑞迪斯先生的參訪非常成功，值得大加讚揚。他們很高興和你們見面，並從參觀我們的工廠中獲益許多。亞歷山大先生指出他提給他公司的報告會是相當正面的。你們同仁對這次參訪的準備是有目共睹的，所以請好好地讚賞自己。

做得好！

肯尼斯

7. **Messrs.** [ˋmɛsɚz] *abbr.* 諸君；各位先生（為 **Mr.** 的複數）
   Messrs. Brown and Johnson will meet you at the airport.
   伯朗先生和強森先生將會和你在機場碰面。

8. **favorable** [ˋfevərəbl̩] *adj.* 贊同的；稱讚的
   The newspaper gave the movie a favorable review.
   這家報紙給這部電影很好的評論。

Language Corner

Praising Someone

Go ahead and give yourselves a pat on the back.

You Might also Say . . .

- You should all pat yourselves on the back for this achievement.
- Congratulate yourselves for a job well done.

B. Confidential Information 機密資訊

Richard 斥責 OEM 代工工廠不應該在 Maples 的客戶來訪時，向其透露產品機密資訊。

TO: Geoff Whiteman
FROM: Richard Barlow
DATE: Aug. 5
SUBJECT: Mr. Small's Visit

Dear Geoff,

Thanks for your help in hosting Mr. Small this past Thursday. I believe the visit went well, **with the exception of**[1] the report you gave him. I'm going to be **blunt**[2] with you, Geoff. That report was a **bombshell**.[3]

I realize you've known Mr. Small for a long time and that you've developed a good working relationship with him. I don't want to **interfere with**[4] this, but you should not have given him GasPack® inspection and life-test data. Some of the information you gave to Mr. Small was confidential and will be **damaging**[5] (like the life-test data that shows we have a serious **durability**[6] problem). What makes the matter even worse is that you never showed the data to us. We didn't know that you have a six-percent life-test defect rate.

Please **keep in mind**[7] that GasPack® technology belongs to Maples and not your factory, and Mr. Small is Maples' GasPack® customer. I hope that in the future you will not give any GasPack® information to anyone without **clearing**[8] it with us first.

We will appreciate your cooperation.

Regards,
Richard

中譯　收件人：吉歐夫・惠特曼

寄件人：理查・巴洛

日期：八月五號

主旨：司默先生的來訪

親愛的吉歐夫：

謝謝你這個星期四幫忙接待司默先生。除了你給他的那份報告之外，我相信這次的參訪很順利。吉歐夫，我要很直接地跟你說：「那份報告非常令人震驚」。

我了解你已經認識司默先生很久了，而你也和他發展了良好的工作關係。我不想要妨礙你們的關係，但你不應該給他 GasPack® 的檢驗和壽命測試的資料。有些你給司默先生的資訊是機密且將帶來損害的（像壽命測試的資料顯示我們有嚴重的耐久性問題）。更糟糕的是你從未讓我們看過這些資料。我們並不知道你的壽命測試有百分之六的不良率。

請記住 GasPack® 的技術屬於 Maples 公司而非你的工廠，而司默先生是 Maples 的 GasPack® 客戶。

我希望未來你不會在沒有我們的認可下，再將 GasPack® 的資料交給其他人。

我們將感謝你的合作。

祝 商安

理查

Vocabulary and Phrases

1. with the exception of   除……以外

With the exception of Carlos, all of you failed to meet the sales goals.

除了卡洛斯之外，你們全部的人都沒有達到業績目標。

2. blunt [blʌnt] *adj.*   直言不諱的；直率的

I'm sorry to be so blunt with you, but we're disappointed in your product.

很抱歉直言不諱地說，但我們對你們的產品很失望。

3. bombshell [ˈbɑmˌʃɛl] *n.*   出乎意料、令人震驚的事

News of Larry being arrested for insider trading was a bombshell to his family.

賴瑞因內線交易而被逮捕的消息令他的家人很震驚。

4. interfere with   妨礙

Please do not interfere with our efforts to start a business.

請不要妨礙我們創業的努力。

5. damaging [ˈdæmɪdʒɪŋ] *adj.*   有害的

He doesn't think the reports will be damaging to our reputation.

他不認為那些報導會有損我們的聲譽。

6. durability [ˌdjʊrəˈbɪlətɪ] *n.*   耐久性

For the past 150 years, this brand of jeans has been known for its durability.

過去的一百五十年，這個牌子的牛仔褲以耐穿而聞名。

7. keep in mind   記住

A good company always keeps its customers in mind.

一個好公司總會將顧客謹記在心。

8. clear [klɪr] *v.*   使獲得批准；認可

I'd love to go out with you, but I need to clear it with my wife first.

我很樂意跟你出去，但是我需要先得到老婆的批准。

Language Corner

Describing Something Unexpected

That report was a bombshell.

You Might also Say . . .

- That report was a huge surprise.
- That report came out of nowhere. What a shocker.
- We never saw that report coming. It was stunning.

C. Still Below Standard 仍未達標準

Stan 向代工生產商反應參觀工廠時所發現的問題。

TO: Jessica Lynch
FROM: Stan Kaufman
DATE: 7/30
SUBJECT: Factory Visit

Dear Jessica,

Thank you for your hospitality during my visit on July 24. James told me I should direct all communication through you, so here are my observations and comments:

1. The clean room was dirty. It was obvious that no regular meaningful cleaning program was **in effect**[1] in this room.

2. One third of the units failed testing after assembly. The assembly procedure must be improved to yield units that are **up to par**.[2]

3. The line people were still not checking the electrode gap with a go / no-go gauge. This could be the cause of lots of the **rejects**.[3]

I know you're new to this project. My frustration is not directed at you, but I must **confess**[4] that I am upset at the lack of cooperation in the factory. Please tell me when these manufacturing **deficiencies**[5] will be corrected.

I must also make known my disappointment with the meeting. I asked a lot of questions but received very few answers. It seemed as though no one was even **remotely**[6] familiar with the production status and problems. Other than personally seeing the lack of progress in production, I consider this trip a waste of my time.

Regards,
eom

中譯 收件人：潔西卡・林區

寄件人：史丹・考夫曼

日期：七月三十號

主旨：參觀工廠

親愛的潔西卡：

感謝妳在我七月二十四號拜訪時熱情的款待。詹姆士告訴我我應該透過妳來作溝通，這是我的看法和意見：

1. 無塵室很髒亂。很明顯地這間工作室並沒有實行定期且有意義的清潔計畫。

2. 組裝完成的組件有三分之一沒有通過測試。組裝的程序必需改進以產出能達到一般水準的組件。

3. 生產線人員仍未使用通過／不通過量規來檢查電極間距。這可能是造成大量瑕疵品的原因。

我知道妳剛接這個案子。我的挫折不是針對妳，但我必需坦白說我對於工廠缺乏合作感到很難過。請告訴我這些製造上的缺陷何時能被改正。

我也必需表明我對這次會議的失望。我問了很多問題，但是幾乎都沒有得到答案。似乎沒有人對生產狀況和問題有一點點的了解。除了看到生產毫無進度之外，我認為此行是在浪費我的時間。

祝 商安

完畢

Vocabulary and Phrases

1. in effect 在實行中；生效

New policies will be in effect as of January 1st.
新政策將會從一月一號開始生效。

2. up to par 達到一般水準

The reason I hired you is that our customer service is not up to par.
我雇用你的理由是我們的顧客服務尚未達到一般水準。

3. reject [ˋriˌdʒɛkt] *n.* 廢棄物；被拒絕的人（物）

The manufacturer has to throw away tons of rejects each year.
那家製造商每年都必需丟棄許多退貨。

4. confess [kənˋfɛs] *v.* 坦白；承認

Willie confessed to breaking the office's coffeemaker.
威力坦承弄壞了辦公室的咖啡機。

5. deficiency [dɪˋfɪʃənsɪ] *n.* 缺點；不足

With your help, I'm sure our company can overcome certain deficiencies.
有你的幫助，我確信我們公司可以克服一些缺點。

6. remotely [rɪˋmotlɪ] *adv.* 極少地；一點點（也）

This is not even remotely close to what they are talking about.
這跟他們所談論的事情一點都不相關。

Language Corner

Complaining about Staff

I must confess that I am upset at the lack of cooperation in the factory.

You Might also Say . . .

- I admit to being disappointed in the lack of teamwork in the plant.
- I must say that I am bothered by the way your staff is working together.
- I'm just going to come out and tell you that I have an issue with your factory operations.

Unit Three
Visa Application / Renewal
簽證申請／更新

不管是出國旅遊或外派都需留意簽證問題，以免因無法入境或逾期居留，而打亂原先的行程規劃。申請簽證或加簽可委託旅行社代辦，如時間緊迫則可親自前往目的國的大使館或辦事處辦理。

✔ Learning Goals

1. 了解辦理工作簽證、居留證等相關程序與應備妥之文件

2. 申請旅行簽證可能會遇到的問題

Before We Start 實戰商務寫作句型

Topic: Asking for and Getting Details 詢問／獲得細節

- (Someone) holds / has a [commercial / marriage / tourist / student / missionary . . .] visa which expires in / is valid (good) until / lasts until . . .
  （某人）持有 [商業／婚姻／旅行／學生／傳教……] 簽證，有效期至……

- (Someone) would like to [apply for / renew / extend / cancel . . .] his (her) visa.
  （某人）想 [申請／更新／延長／取消……] 他（她）的簽證。

- Would you please tell me if [you foresee / you're expecting / I may encounter / there could be . . .] any problems in + V-ing?
  可以請你告訴我在……方面是否 [你預見／你預期／我有可能遇到／可能有……] 任何問題？

- What's the [usual / normal / expected . . .] waiting period after filing / putting in / completing an application?
  提出申請後 [通常／預計……] 要等多久？

- We'll need [a copy of your university degree / several letters of reference / a tax statement . . .].
  我們將需要 [你大學文憑的副本／幾封推薦信／報稅證明書……]。

- This [e-mail / document / information . . .] is not for general distribution / circulation / dispersal.
  請勿公開傳閱／散布這封（份）[電子郵件／文件／資訊……]。

- I'm soliciting / seeking / asking for / requesting / calling for your help / aid / assistance.
  我請求你的幫忙。

A. Renewing a Visa 更新簽證

主管要到印尼更新商業簽證,於是秘書 **Pamela** 先寫信詢問駐在國的代表辦事處相關的申請手續。

To: Taipei Economic and Trade Office, Jakarta
Cc: Adam Holt
From: Pamela Wang
Date: November 20
Subject: Visa Requirements

Dear Sir / Madam,

My employer, Mr. Adam Holt, is an American **residing**[2] in Taiwan. Right now, he holds a **commercial**[3] visa which **expires**[4] in early December. He would like to **vacation**[5] in Bali and while he's there renew his visa. Would you please tell me if you foresee any problems in him getting a new one during his trip? He is not a permanent resident of Taiwan yet. He is, however, married to a Taiwanese woman.

Once in Bali, my employer plans to fly to Jakarta to visit your office concerning the visa. What's the usual waiting period after **filing**[6] an application? If there is indeed a certain amount of time, I'd like to know the following: is there any way that we can start the process from here in Taiwan? I'd also like to find out what the fees will be and any other **relevant**[7] information. Please remember that expedience is of the essence.

Your help in this matter is very much appreciated. I look forward to your response.

Best regards,
Pamela Wang
Secretary, Adam Holt Design Partners

中譯　收件人:駐雅加達台北經濟貿易代表處
　　　副本抄送:亞當‧豪特
　　　寄件人:王潘蜜拉
　　　日期:十一月二十號
　　　主旨:簽證規定

親愛的先生 / 女士:

　　我的雇主豪特‧亞當先生,是一位住在台灣的美國人。現在他持有的商業簽證十二月初就到期了。他想要去峇里島度假,並在那裡更新他的簽證。可以請您告訴我他此行在申請新簽證方面您是否有預見任何問題?他還不是台灣的永久居民,但他娶了台灣人。

　　到了峇里島,我的雇主計劃飛往雅加達到貴辦事處辦理簽證事宜。請問提出申請後通常要等多久?如果的確需要一段時間,我想了解下列事項:我們是否可以在台灣就開始進行申請的程序?我也想知道簽證費用以及其他相關的資訊。請記住最重要的是要方便。

非常感謝您對此事的幫忙。我期待您的回信。

謹上

王潘蜜拉

亞當‧豪特設計事務所 秘書

Vocabulary and Phrases

1. renew [rɪˋnju] *v.* （使）更新；重新開始（renewal *n.*）

If you use our express service, we can renew your passport in 24 hours.

如果你使用我們的快速服務，我們可以在二十四小時內更新你的護照。

2. reside [rɪˋzaɪd] *v.* 居住；駐在

How long have you resided in Hong Kong?

你住在香港多久了？

3. commercial [kəˋmɝʃəl] *adj.* 商業的；營利本位的

Mike has a commercial pilot license.

麥克擁有商用機師執照。

4. expire [ɪkˋspaɪr] *v.* 到期；期滿

We can't issue you a visa because your passport has expired.

因為你的護照過期了，所以我們不能核發簽證給你。

5. vacation [veˋkeʃən] *v.* 度假

The Gunderson's are vacationing in Florida this year.

甘德森家族今年要去佛羅里達度假。

6. file [faɪl] *v.* 提出（申請）；提起（訴訟）

The reason Mrs. Cooke is filing for a divorce is that her husband went bankrupt.

庫克太太因她老公破產而正在申請離婚。

7. relevant [ˋrɛləvənt] *adj.* 有關的

Why are you talking about that project? It's not relevant to our discussion.

你為什麼談到那個案子？它跟我們的討論無關。

Language Corner

This Is Important

+ be of the essence
+ be of the utmost importance
+ be of much significance
+ be vital / crucial

For example:

1. Mr. Walters needs the report today — time is of the utmost importance.

2. You have an hour to make this delivery. Speed is crucial.

外籍員工 Tony 要申請居留簽證，人事部的 Carrie 和他確認需要準備的文件。

○ ○ ○

TO: Tony
FROM: Carrie
DATE: August 21
SUBJECT: Processing Your Paperwork
ATTACHMENT: We Still Need these Items

Dear Tony,

We're still **putting the final touches on**[1] your visa application. But we need a couple of things on your side. Please see the list I attached. We're dealing with time **constraints**,[2] so please take care of these points as soon as possible.

Cheers,
Carrie

Attachment

1. A letter from your previous employer that shows you've worked in the industry.
2. A copy of your university degree. Please make sure the name exactly matches that on your passport. If your passport shows your middle name written in full, your university degree must do so **likewise**.[3]
3. Other **certification**[4] that you think might be useful.

TO: Carrie
FROM: Tony
DATE: August 22
SUBJECT: Re: Processing Your Paperwork

Hi Carrie,

I just **fired off**[5] a copy of my degree. It's already been **notarized**[6] and stamped by the Taipei Economic and Cultural Office in LA. The degree is not an **original**,[7] so you can keep it. BTW, I've sent you three letters of reference. They should be with my resume. Do you need me to send them again? And what kind of certification are you talking about?

Here's some general information you may need:

Tony Timothy Bowen

Passport no.: DR329882

DOB: 5/10/57, in Long Beach, CA

Last residence in the US: 56 Park Road, Scotia, IL

Spouse:[8] Mary Bowen, Taiwan citizen

Please let me know soon if you need anything further. Thanks, once again, for all of the assistance you've provided.

THIS E-MAIL IS NOT FOR GENERAL DISTRIBUTION. THANK YOU.

中譯

收件人：東尼
寄件人：凱莉
日期：八月二十一號
主旨：處理你的文件
附件：我們仍需要這些項目

親愛的東尼：

我們仍在處理你簽證申請的最後一些事項。但是我們需要你提供一些東西。請看我附上的清單。我們有時間上的限制，所以請儘快處理這些事。

工作愉快
凱莉

 附件

1. 一封由你前雇主證明你在這行工作過的信函。

2. 一份大學文憑的副本。請確認名字與護照上的完全相符。如果你護照上的中間名有全部拼出來，你的大學文憑也必需如此。

3. 其他你認為有幫助的證明文件。

收件人：凱莉
寄件人：東尼
日期：八月二十二號
主旨：回覆：處理你的文件

凱莉，妳好：

我剛寄了一份我的學歷影本。它已經由駐洛杉磯台北經濟文化辦事處公證並蓋章。這張文憑並非原稿，所以妳可以把它留著。對了，我寄了三封推薦信給妳，它們應該和我的履歷表放在一起，妳需要我再寄一次嗎？另外，妳是指什麼樣的證明？

這裡有一些妳可能會需要的一般性資訊：

東尼・提摩西・伯恩
護照號碼：DR329882
生日：1957 年十月五號（加州長灘市）
在美國最後的住所：伊利諾州斯科舍市公園路 56 號
配偶：瑪莉・伯恩（台灣公民）

如果妳需要任何進一步的資訊，請儘速讓我知道。再一次感謝妳提供的所有協助。

請勿公開傳閱這封電子郵件。謝謝妳。

Vocabulary and Phrases

1. put the final touches on
完成……的最後一部分工作；對……作最後潤飾
I'm still putting the final touches on this bid. It's almost finished.
我還在處理這次投標的收尾工作。差不多要完成了。

2. constraint [kən'strent] *n.* 限制；約束
As our department is facing serious financial constraints, we're unable to hire anyone else.
由於我們部門面臨嚴重的財務限制，我們無法再聘請其他人了。

3. likewise ['laɪkˌwaɪz] *adv.* 同樣地；照樣地
We're making copies of the contract and suggest that you do likewise.
我們在影印合約的副本，我們也建議你這麼做。

4. certification [ˌsɝtəfəˈkeʃən] *n.* 證書；檢定
While on vacation, Danny managed to get Open Water scuba diver certification.
度假期間丹尼設法取得開放水域潛水員的證書。

5. fire off （口語）寄送
Could you fire off this letter while you're at the post office?
你去郵局的時候可以請你寄一下這封信嗎？

6. notarize ['notəˌraɪz] *v.* 公證
All documents for a marriage visa must be notarized before you hand them in.
所有結婚簽證的文件在提交之前都必需公證。

7. original [əˈrɪdʒənl̩] *n.* 原稿；原著
Here's a copy of my birth certificate. I don't have the original anymore.
這是我出生證明的副本。我沒有正本了。

8. spouse [spauz] *n.* 配偶
Although my spouse is Taiwanese, she didn't keep her maiden name.
雖然我的配偶是台灣人，但是她沒有保留她的娘家姓。

Language Corner

No Time to Waste

We're dealing with time constraints, so please take care of these points as soon as possible.

You Might also Say . . .

- Time is tight, so please deal with these issues as quickly as you can.
- We don't have any time to spare. Please get on top of this ASAP.

Carl 和太太計畫到歐洲度假，但卻碰到一些簽證問題。

TO: Jacques Brodeur
FROM: Carl Peng
DATE: 1/4
SUBJECT: Can You Help Us Out?

Dear Jacques,

We've run into a snag in our plan to go to Europe. Right now, I'm **soliciting**[2] your help. Mary has been collecting visas for the countries we're planning to visit or travel through, and she's just discovered that France stopped **issuing**[3] visas to Taiwanese individuals, unless the applicants can show that they're going to France for business purposes. The French visa office here in Taipei will accept a fax from a French company which **extends**[4] an invitation to visit. It needn't be detailed or **verbose**.[5]

Could you call one of your contacts in France and ask him or her to send a fax to me, saying "looking forward to you and your wife visiting our factory next month" **or some such stuff**?[6] And if you can't or your contact isn't around, would you please ask Henri Potvin to do it? We**'re running short on**[7] time and we have other visas to get. Originally, Paris was going to be our point of entry. We hope we can stick to the plan.

If you can **pull** this **off**,[8] we'll be most indebted.

Sincerely,
Carl

中譯 收件人：賈克・布羅杜爾
寄件人：彭卡爾
日期：一月四號
主旨：你可以幫我們忙嗎？

親愛的賈克：

我們要去歐洲的計畫突然遇到了一點麻煩。現在我要請求你的幫忙。瑪莉正在收集我們打算造訪或經過的國家的簽證，而她剛發現法國已經停止核發簽證給台灣個人旅行者，除非申請人可以證明他們是去法國經商。在台北的法國簽證辦事處可以接受來自法國公司邀請拜訪的傳真，內容不需太詳細或很冗長。

可以請你打電話給你在法國的熟人，並請他（她）傳一份上面寫著「期待您和夫人下個月拜訪我們的工廠」或諸如此類的傳真給我嗎？如果你沒有辦法或你的熟人不在的話，可以麻煩你請亨利・普托文來處理這件事嗎？我們快沒時間了，而且還有其他的簽證要申請。原先我們要從巴黎進入歐洲。我們希望能按照原訂計畫進行。

如果你能完成這項艱難的任務，我們將萬分感激。

祝 商安
卡爾

Vocabulary and Phrases

1. complication [ˌkɑmpləˈkeʃən] *n.*
（突然出現的）困難；障礙

This deal isn't going so well. In fact, there are several complications.
這筆交易進行地不太順利。事實上，有一些困難。

2. solicit [səˈlɪsɪt] *v.* 請求；懇求

Any government official found to be soliciting gifts will be fired.
任何政府官員被發現要求禮物將會被革職。

3. issue [ˈɪʃju] *v.* 核發；發行

Where was your passport issued?
你的護照是在哪裡核發的？

4. extend [ɪkˈstɛnd] *v.* 給予；提供

Let's extend a warm welcome to Ms. Porter.
讓我們熱情歡迎波特女士。

5. verbose [vɚˈbos] *adj.* 冗長的；嘮叨的

This article is too verbose. Please shorten it.
這篇文章太冗長了。請把它縮短。

6. or some such stuff 或諸如此類

I think the company sells computers, cell phones, or some such stuff.
我想那家公司是在賣電腦、手機或諸如此類的東西。

7. be running short on 快用完……

After being stuck in traffic for three hours, Cliff was running short on patience.
在被困在車陣中三小時後，克里夫快要失去耐性了。

8. pull sth off 成功地完成某件難事

He doesn't think they'll be able to pull that off. The order is simply too big.
他不認為他們可以成功地完成那個艱難的任務。那個訂單真的太大了。

Language Corner

Getting It Done

If you can pull this off, we'll be most indebted.

You Might also Say . . .

- If you can manage this, it would be greatly appreciated.
- If you could get this to work, we'd be most thankful.
- If you can handle this, it'll be a big load off our shoulders.

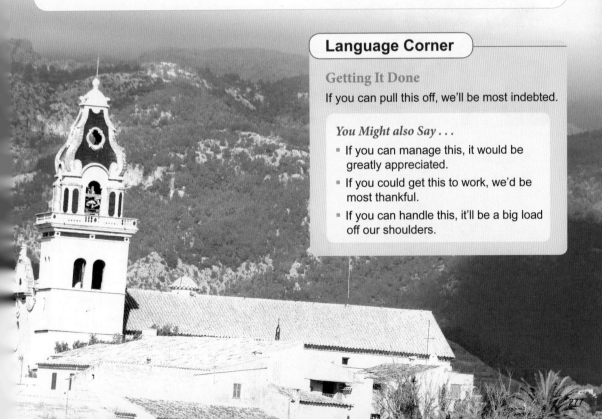

Unit Four
Meeting Minutes
會議記錄

會議記錄不一定要由秘書來做，負責記錄的人通常不是會議的主要參與者，但其對所要討論的議題很了解，能將與會者的報告、開會決議、待辦事項、任務分配等快速地作摘要，並於會後整理成正式的會議記錄。

✅ Learning Goals

1. 了解會議記錄需包含哪些內容

2. 會議記錄務求精簡，不可過於瑣碎或加油添醋

3. 記錄經主席過目後即可分發給全體與會者，以確保決議事項能被確實執行

Before We Start 用英語開會

Topic 1: 會議記錄撰寫守則

英文會議記錄並無固定的形式，但通常會包含下列元素：

- 日期（date）、時間（time）和地點（location）
- 召開會議的目的（purpose / objective / goal）
- 與會人員（participants / attendees / present）

  在出席人員名單上亦會加註他們的頭銜，如主席（chair / chairman / chairperson）、顧問（adviser）、代理人（proxy）等。
- 因故未能出席的名單（absence）
- 會議中達成的決議（decision / resolution）
- 是否有提出動議（motion）或其他討論事項（other business）
- 下次會議時間、地點及討論事項（agenda）

Topic 2: 與開會相關的英文用語

- **The meeting is adjourned until next Friday.**
  會議延至下星期五。

- **It was decided by a unanimous vote that the new factory should be built.**
  一致投票通過決議要興建新的工廠。

- **I move to table / shelve / put off the [motion / offer / project . . .].**
  我提議擱置 / 暫緩那項 [動議 / 提案 / 計畫⋯⋯]。

- **The budget proposal was approved / rejected by a vote of 20 to 12.**
  以二十對十二票通過 / 否決該項預算案。

- **The board ratified the investment, with six votes in favor, two against, and one abstention.**
  董事會批准該項投資，有六票贊成、兩票反對和一票棄權。

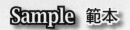

From: Tina Hsiao
To: James Sun; Patty Lo
Cc: Yolanda; Ben Maples; Jack Huang; Richard Barlow
Sent: Wednesday, August 30 4:15 p.m.
Subject: Meeting Minutes
Attachment: minutes8_30.doc

Dear Everyone,

Yesterday's meeting went extremely well. Please see the attached minutes.

Respectfully submitted,
Tina Hsiao
Assistant Project Manager
Maples Corporation

 Attachment

Attendees:[1] Ben Maples (Chair), Jack Huang, Richard Barlow — Maples Corporation
James Sun, Patty Lo — Global OEM
Yolanda Ho — Global OEM (lunch only)

On August 29, the above-listed people from Maples Corporation and Global OEM met in the Marco Polo Hotel to discuss the past, current, and future status of GasPack® production in China.

Ms. Ho joined the group for lunch after the meeting was **concluded**.[2] The highlights of the meeting were as follows:

1. Mr. Sun informed everyone that Ms. Ho would not be able to attend the meeting. She would, however, be joining the group for lunch.

2. Mr. Huang briefly ran over resolutions from last month's meeting.

3. Mr. Sun reported that the current maximum capacity of the production line was 2,400 units per day, with a future capability of 7,200 units per day.

4. In response to concerns voiced by some of the MC people that there was not a responsible person in charge of its line in China, GO promised to assign Mr. Yang to be the full-time production **foreman**[3] of the GasPack® line. Mr. Yang will report directly to C. C. Lee, the factory production manager.

5. GO promised to **formulate**[4] new IQC inspection procedures for the GasPack® line in the next few weeks.

6. Ben Maples told GO that the minimum shipping requirements for the balance of this year total 180K units. GO and MC agreed to produce 60K units per month from now until the end of the year, at which time GO should be able to begin manufacturing 100K per month.

7. Ben Maples agreed to supply GO with updated three-month forecasts, by customer, on an **ongoing**[5] basis.

8. GO agreed to continue selling the units to MC at the current prices, except for the #2C and #5P units, which will have **substantiated**[6] cost increases resulting from packaging, printing, or component **surcharges**.[7] MC agreed.

9. Over lunch, Ms. Ho told us that the new QC manager, Peter Liang, will report directly to her. She has given Mr. Liang the budget and power necessary to build the GO QC department into a strong, effective group.

The meeting was a positive one, with both sides indicating a desire to continue promoting the business together.

Recorded by: Tina Hsiao
Assistant Project Manager

Confirmed by: Ben Maples
Chair

中譯　寄件人：蕭蒂娜

收件人：孫詹姆士；羅佩蒂

副本抄送：尤蘭妲；班‧梅普爾斯；黃傑克；理查‧巴洛

傳送時間：八月三十號星期三下午四點十五分

主旨：會議記錄

附件：minutes8_30.doc

親愛的各位：

昨天的會議進行得非常順利。請參閱附件的會議記錄。

謹呈
蕭蒂娜
專案副理
Maples 公司

 附件

與會人員：班‧梅普爾斯（主席）、黃傑克、理查‧巴洛 — Maples 公司

孫詹姆士、羅佩蒂 — Global OEM

何尤蘭妲 — Global OEM（僅參加午餐）

上述來自 Maples 公司及 Global OEM 的人員於八月二十九號在馬可波羅飯店會面，討論 GasPack® 產品在大陸過去、現在和未來的生產情況。

會議結束後，何女士和大家共進午餐。會議重點如下：

1. 孫先生通知大家何女士將無法出席會議，但她在午餐時會加入。

2. 黃先生很快地簡述上個月會議的決議。

3. 孫先生報告目前生產線最大的產量為每天兩千四百件，未來可達每天七千兩百件。

4. 回應部分 MC 人員擔心沒有人負責他們在大陸的生產線，GO 答應指派楊先生擔任 GasPack® 生產線的全職領班。楊先生將直接向工廠生產經理李 C.C. 報告。

5. GO 承諾在接下來幾週會針對 GasPack® 生產線制定新的進貨品管檢驗程序。

6. 班‧梅普爾斯告訴 GO 今年剩餘月分的最低出貨需求總計為十八萬件。GO 和 MC 同意從現在起到年底每個月生產六萬件，屆時 GO 每個月應可開始生產十萬件。

7. 班‧梅普爾斯同意繼續提供 GO 每個客戶最新三個月的（採購）預測。

8. GO 同意繼續以現在的價格販售產品給 MC，但不包括 #2C 和 #5P 產品，其將因包裝、印刷或零件漲價而導致價錢有實質的增加。MC 表示同意。

9. 午餐期間，何女士告訴我們新的品管經理梁彼得將直接向她報告。她已經給予梁先生所需的預算和權力來將 GO 的品管部門發展成強而有力的單位。

這次的會議很成功，雙方均表明欲繼續共同促進生意的意願。

記錄：蕭蒂娜
專案副理

覆核：班‧梅普爾斯
主席

Vocabulary and Phrases

1. attendee [əˌtɛnˋdi] *n.* 出席者；在場者

There were over 100 attendees at the banquet.

這場宴會有超過一百人出席。

2. conclude [kənˋklud] *v.* （使）結束

The workshop will conclude at four o'clock.

那場研討會將在四點結束。

3. foreman [ˋfɔrmən] *n.* 領班；工頭

John's father was a foreman at the sawmill.

約翰的父親以前是鋸木廠的領班。

4. formulate [ˋfɔrmjəˌlet] *v.*

規劃（制度等）；想出（計畫等）

They are formulating fresh ideas to deal with this issue.

他們正在想新的點子來解決這個問題。

5. ongoing [ˋɑnˌgoɪŋ] *adj.* 不間斷的；進行中的

In the past, we had an ongoing agreement that prices would remain the same.

過去我們一直有個價錢不變的協議。

6. substantiated [səbˋstænʃɪˌetɪd] *adj.*

經證實的；有根據的

You can only get your payment if your claims are substantiated.

索賠必需經證實，你才能拿到錢。

7. surcharge [ˋsɝˌtʃɑrdʒ] *n.* 漲價；額外費用

Is this simply the charge for the product or does it include the surcharges as well?

這只是產品的價格還是也包含了額外費用？

Chapter Eight

Protecting the Company's Interests

Unit One
Patent Oppositions 專利異議

具有創新設計的產品在上市前一定要在產地和銷售市場所在的國家申請專利，如果讓他人先取得專利，其有可能透過法律程序禁止別家公司出貨和販賣。但不可能在每個國家都申請專利，最主要是要設法取得台灣、大陸、美國、日本及歐洲等主要地區的專利權。

✓ Learning Goals

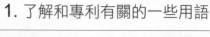

1. 了解和專利有關的一些用語

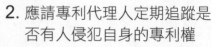

2. 應請專利代理人定期追蹤是否有人侵犯自身的專利權

3. 採取法律行動前需先發警告信或存證信函給侵權的對方

Before We Start 實戰商務寫作句型

Topic: Dealing with Patent Issues 處理專利權事宜

- He will alert us to / warn us of any patent applications filed for anything related to / connected to / having to do with our product line.
  他將會提醒我們注意任何與我們產品線相關的專利權申請。

- Do we have any patents granted / issued / in place / in effect in Taiwan?
  我們在台灣有（被授予 / 核發）任何專利權嗎？

- We hold / believe / think that your patent blatantly / obviously / clearly infringes on (upon) / encroaches on (upon) / breaches . . .
  我們認為你的專利公然 / 明顯地侵犯……

- They deemed features / characteristics / attributes of our product to be unique / distinctive / peculiar to the other company's design.
  他們認為我們產品的特色（性）是另一家公司獨有的設計。

- We can successfully appeal the [loss / decision / ruling . . .].
  我們可以成功地對此 [敗訴 / 決定 / 裁決……] 提出上訴。

- Shall we proceed / go ahead / move forward / take the next step?
  我們應該繼續進行 / 採取下個步驟嗎？

A. Checking Patents 核對專利權

Maples 公司的專利權受到侵犯，Richard 乃委託專利代理人提出異議並向總公司詢問在台灣的專利申請狀況。

TO: Tony Hughes
FROM: Richard Barlow
DATE: 7/1
SUBJECT: Patents

Dear Tony,

　　We've hired a patent attorney named Mr. Lin. He's an engineer who is government-**certified**[1] to file and oppose patents and trademarks, and was **referred**[2] to us because he has gas technology experience. Mr. Lin will file an opposition to the two patents that **infringe upon**[3] ours. It will speed things up considerably if we supply him with our patent paperwork. Would you FedEx me a copy of all of our patents?

　　In the future, Mr. Lin will alert us to any patent applications filed for anything related to our product line. We estimated that, **compared to**[4] going through an **intermediary**,[5] having Mr. Lin on board to do the patent work would save us about two-thirds of the expense.

　　Lastly, do you know if we do, in fact, have any patents granted in Taiwan? Ted has informed me that we do, but I can find no **documentation**[6] of any kind that **attests to**[7] this. Mr. Lin will do his own search for our patents in Taiwan. If you already know for sure whether or not we have any, it will help speed up the process. Thanks.

Take care,
Richard

中譯　收件人：東尼‧休斯

寄件人：理查‧巴洛

日期：七月一號

主旨：專利權

親愛的東尼：

　　我們雇用了一位專利代理人 — 林先生。他是一名獲得政府認證可對專利權和商標提出申請和異議的工程師。他之所以被引薦給我們是因為他有瓦斯技術的經驗。林先生將會對兩項侵犯到我們權益的專利權提出異議。如果我們可以提供他我們專利權的相關文件，將會大大加速此事的進行。你可以用快遞寄一份我們所有專利權的副本給我嗎？

未來林先生會提醒我們注意任何與我們產品線相關的專利權申請。我們評估過相較於透過中間人，雇用林先生來處理專利權事宜可為我們省下大約三分之二的開銷。

最後，你知道我們在台灣實際上真的有被授予任何專利嗎？泰德告訴我我們有，但是我找不到任何文件來證明。林先生將會自行搜尋我們在台灣的專利權。如果你已經確定知道我們是否擁有任何專利，將有助於流程加速進行。謝謝。

保重

理查

Vocabulary and Phrases

1. certify [ˋsɝtəˌfaɪ] *v.* 發證書（或執照等）給

The priest was certified to perform marriages in Texas.
這位神父領有執照可在德州證婚。

2. refer [rɪˋfɝ] *v.* 引薦；使求助於

An ex-colleague referred the certified public accountant (CPA) to us.
之前的同事向我們推薦這位註冊會計師。

3. infringe upon (on) 侵犯；侵害

Our competitor's new product infringes upon our patent.
我們競爭對手的新產品侵害到我們的專利權。

4. compared to 相較於

Buying food at the grocery store is cheap compared to eating out.
相較於外食，在雜貨店買食物算是便宜的。

5. intermediary [ˌɪntɚˋmidɪˌɛrɪ] *n.* 中間人；媒介

I couldn't contact the mayor, so I left him a message through an intermediary.
我聯絡不上市長，所以我透過一名中間人留了信息給他。

6. documentation [ˌdɑkjəmɛnˋteʃən] *n.*
（總稱）文件

You need proper documentation to file a claim.
你需要適當的文件來申請索賠。

7. attest to 證實；證明

I can attest to the fact that the food at this restaurant is delicious.
我可以證明這間餐廳的食物很美味。

Language Corner

Getting Things Done Promptly

It will speed things up considerably if we supply him with our patent paperwork.

You Might also Say . . .

- It will expedite things substantially if we give him our patent paperwork.
- It will get things moving if we provide him with documents for our patents.
- It will quicken the process if we hand over our patent papers to him.

B. Notification of an Opposition 通知將提出專利異議

Sun Melding Supplies 公司的專利權遭人侵犯，在提出異議申請之前，先委託律師發函要求對方停止相關產品之生產、出貨及販售。

September 13

Legal Department
Ironworks, Inc.
120 Zhongzheng Rd., Suite 967
Hsinchu, Taiwan, ROC

To Whom It May Concern:

I'm writing this letter on behalf of Sun Melding Supplies to forewarn you that on Wednesday, September 27, the company will be filing an opposition to the patent you recently **acquired**[1] for a gas-powered hand tool.

Attached are copies of the hand-tool patents of both your company, Ironworks, Inc., and Sun Melding Supplies. We hold that your patent **blatantly**[2] infringes on one already held by Sun Melding Supplies. It is in your company's best interests to stop manufacturing, shipping, and marketing the products while we are filing a patent opposition at the Taiwan Intellectual Property Office. Once the opposition is approved, we will, if necessary, proceed with legal action to have your patent **retracted**.[3]

Contact me at any time should you have any questions concerning the matter.

Sincerely,
Johnson Wu
Patent Attorney
Sun Melding Supplies

中譯

九月十三號

法務部門
Ironworks, Inc.
台灣省新竹市
中正路 120 號 967 室

敬啟者：

我謹代表 Sun Melding Supplies 寫這封信事先告知你們，該公司將於九月二十七號星期三針對你們最近獲得的一項瓦斯手動工具的專利權提出異議。

附件為你們公司（Ironworks, Inc.）和 Sun Melding Supplies 手動工具專利權的文件副本。我們認為你們的專利權公然侵犯 Sun Melding Supplies 已經擁有的一項專利之權益。在我們向台灣智慧財產局提出專利異議申請的同時，停止相關產品之生產、出貨及販售是對貴公司最有利的。一旦異議申請獲准，如有必要我們將會採取法律行動撤銷你們的專利。

如果你們對此事有任何問題，請隨時與我聯絡。

祝 商安
吳強森
專利代理人
Sun Melding Supplies

C. Planning an Appeal 計劃上訴

異議申請被駁回後 Sean 詢問總公司是否要提出上訴。

TO: Harold Reynolds
FROM: Sean Latimer
DATE: 3/19
SUBJECT: Failed Opposition

Harold,

We lost our opposition to the hand-tool patent that had been issued to a company in Hsinchu, Taiwan. In my opinion, the reasons the Taiwan Intellectual Property Office listed for refusing our opposition don't make sense. They deemed features of our product (such as the ignition system and the device designed for mixing air and gas) to be unique to the other company's design. Apparently, they didn't even look at our patent.

Our patent attorney thinks we can successfully **appeal**[4] the loss (an appeal costs around US$600). Shall we proceed? As it stands now, this other company could seek an **injunction**[5] against our hand-tool shipments until we have the matter **sorted out**[6] in court.

Best,
Sean

中譯

收件人：哈洛德‧雷諾茲
寄件人：尚恩‧拉提默
日期：三月十九號
主旨：異議無效

哈洛德：

我們針對核發給台灣新竹市一家公司的手動工具專利權所提出的異議被駁回。我覺得台灣智慧財產局所列出拒絕我們異議的理由並不合理。他們認為我們產品的特色（如點火系統和設計用來混合空氣和瓦斯的裝置）是另一家公司獨有的設計。顯然他們甚至沒有看我們的專利權。

我們的專利代理人認為我們可以成功地對此敗訴提出上訴（上訴費用約為六百元美金）。我們應該繼續進行下去嗎？依目前的情況看來，直到我們於法院解決此事之前，對方公司可以對我們手動工具的出貨申請禁制令。

祝 一切安好
尚恩

1. acquire [əˈkwaɪr] *v.* 取得；獲得

Mr. Kelly acquired his degree from a prestigious university in the US.
凱利先生於美國一所知名的大學取得他的學位。

2. blatantly [ˈbletn̩tlɪ] *adv.* 公然地

The author blatantly copied another person's work.
那位作者公然抄襲另一個人的作品。

3. retract [rɪˈtrækt] *v.* 撤銷；收回

It would be smart to retract your inappropriate statement immediately.
立即撤回你們不恰當的聲明是明智的。

4. appeal [əˈpil] *v.* 對……提出上訴

The man planned to appeal the judge's decision.
那名男子計劃對法官的判決提出上訴。

5. injunction [ɪnˈdʒʌŋkʃən] *n.*
（法院的）禁止令；強制令

We're going to seek an injunction to stop them from producing that product.
我們要向法院申請禁制令，阻止他們生產那項產品。

6. sort out 解決

She's going to get our fund shortage sorted out by the end of the month.
她將在月底前解決我們資金不足的問題。

➤ Language Corner

Giving Advice

+ It is in your best interests . . .
+ It would serve you best . . .
+ The best thing you can do is . . .
+ It would be highly favorable to you . . .
   ……對某人最有利。

For example:

1. It is in your best interests to comply with the agreement.
2. It would serve you best to take this offer.

Notes

Unit Two
Confidentiality / Nondisclosure Agreements
保密合約

進行商業合作計畫時，為防止某方洩露機密資訊，通常需簽署保密協議。本篇範例是 H&T Distributors LLC 公司請 OEM 工廠針對要量產的新產品來報價，但因該產品的專利尚未獲准，為防止產品設計的機密外洩，故要求工廠簽署保密合約。

✔ Learning Goals

CONFIDEN

1. 了解保密協議的格式及寫法

2. 保密協議為有效法律文件，遣詞用字需十分明確以免引發爭議

3. 保密協議需於收信人簽名回傳之後才算生效

○ ○ ○

Date: Sept. 1
ADDRESSEE:[1] Global Appliances Limited
            610 W. Vine Street
            Kissimmee, Florida 34744
Attention: Mr. Michael Yang
            Engineering Department

This letter concerns a proposed **disclosure**[2] that we wish to make to Global Appliances Limited (hereafter "the **Recipient**"[3]) relating to an automatic water control device design for kitchens and other use (hereafter "the AWCD Information") as well.

We consider the AWCD Information to be proprietary and of considerable commercial value. Therefore, we are willing to disclose the AWCD Information to the Recipient upon condition that the Recipient accepts the information **in confidence**,[4] makes no use of the information other than for purposes **set forth**[5] **herein**,[6] and takes **safeguards**[7] to maintain the confidentiality of the AWCD Information no less stringent than those used by the Recipient to protect its own confidential information.

The AWCD Information shall include information disclosed to the Recipient in a letter marked confidential and / or **orally**[8] disclosed to the Recipient provided that the oral disclosure is reduced to a letter in a summary form within 30 days of disclosure and submitted to the Recipient with a confidential marking.

The AWCD Information disclosed to the Recipient needs not to be treated as confidential if any of the following applies:

a. Information was **in the possession of**[9] the Recipient prior to disclosure to the Recipient from H&T Distributors LLC.

b. Information becomes a part of the **public domain**[10] as a consequence of publication, patenting, or otherwise through no act or fault of the Recipient.

c. Information is disclosed to the Recipient by a third party that has been granted the right to make such a disclosure.

The term of this Agreement and the time period during which confidential information will (may) be disclosed hereunder shall be two years from the date of acceptance of this letter. The period of confidentiality shall be during the term of this Agreement and for a period of two years subsequent to its expiration after which time the burden of confidentiality shall **terminate**.[11]

No rights to the AWCD Information are granted to the Recipient, **express or implied**,[12] other than the right to use the AWCD Information for purposes set forth herein.

If the above is acceptable, please acknowledge acceptance on behalf of the Recipient by signing the enclosed copy of this letter and returning the signed copy to me at your earliest convenience.

Offered by:
COMPANY NAME: H&T Distributors LLC
By: *Grey Little*
General Manager
Date: *Sept. 1*

Accepted by:
COMPANY NAME: Global Appliances Limited
By: *Michael Yang*
Engineering Manager
Date: *Sept. 8*

中譯

日期：九月一號

收件人：34744 佛羅里達州基西米市

樹藤西街 610 號

Global Appliances Limited

工程部 楊麥克先生 收

這封信是關於我們想要向 Global Appliances Limited（此後稱「接受者」）有計畫地公開有關一項針對廚房和其他用途的自動控水裝置設計（此後稱「AWCD 資訊」）。

我們認為 AWCD 資訊是我們的專利且具有相當的商業價值。因此，假如接收者能保密地接受訊息，除了下文所載明的情形外不將其用於他途，並採取如同保護自身機密資訊般嚴屬的防護措施，我們願意向接受者公開 AWCD 資訊。

AWCD 資訊應包含以標明為機密的信函傳送給接受者的資訊，和（或）用口述的方式向接受者披露，唯公開的內容需於三十天內被簡化為摘要，且以註記為機密的信函提交予接受者。

向接受者公開的 AWCD 資訊在下列任一種的情況下無需被視為機密：

a. 在 H&T Distributors LLC 向接受者揭露此資訊前，接受者已擁有該項訊息。

b. 由於刊物出版、專利取得，或其他非接受者的行為或失誤所導致的原因，而使該資訊為或變成公共領域的一部分。

c. 資訊是由擁有權利公開訊息的第三者所揭露給接受者的。

這份合約的期限與依此將（可以）公開機密資訊的有效期為自接受這份信函後的兩年。合約期間與約滿後的兩年皆為保密期，在那之後保密的責任才能終止。

除了於此載明的 AWCD 資訊使用權限之外，無明示或暗示授予接受者使用該項資訊的權利。

如可接受上述內容，請簽署附上的信函副本，方便的話請將已簽署的副本儘早寄回給我，以表示代表接受者同意。

提供者：

公司名稱：H&T Distributors LLC

葛雷‧理特

總經理

日期：九月一號

接受者：

公司名稱：Global Appliances Limited

楊麥克

工程部經理

日期：九月八號

Vocabulary and Phrases

1. addressee [ˌædrɛˋsi] *n.* 收信人；收件人

Shannon has already had those items delivered to the addressee.

雪儂已經把那些物品寄給對方了。

2. disclosure [dɪsˋkloʒɚ] *n.* 揭露；公開（disclose *v.*）

Doctors shouldn't make disclosure about patients' personal medical information.

醫生不應公開病人私人的就醫資訊。

3. recipient [rɪˋsɪpɪənt] *n.* 接受者；受領者

If I don't find a job soon, I'm going to have to become a welfare recipient.

如果我沒有馬上找到工作，我就要領社會救濟金了。

4. in confidence 保密地

Please keep in mind that this information has been given to you in confidence.

請記住提供您的這份資訊需保密。

5. set forth 提出；說明

All of the terms set forth in this contract must be honored.

這份合約提出的所有條件都必需被實踐。

6. herein [ˌhɪrˋɪn] *adv.* 此中；於此

I, Barry Oberly, herein agree to abide by these rules.

我（貝瑞‧歐博禮）於此同意遵守這些規定。

7. safeguard [ˋsɛfˌgɑrd] *n.* 預防措施；保護

As a safeguard, Mr. Plante had his secretary make backup files.

作為預防措施，布藍德先生要他的秘書將文件做備份。

8. orally [ˋɔrəlɪ] *adv.* 口頭上

Before people could write, history was passed down orally.

在人們會書寫之前，歷史是經由口述傳承的。

9. in the possession of 為……所有

Any employee caught in the possession of a firearm will be immediately dismissed.

任何被發現持有槍枝的員工會馬上被解雇。

10. public domain 公共領域

After the copyright expires, the work will enter the public domain.

在著作權期滿後，該作品將屬於公共領域。

11. terminate [ˋtɝməˌnet] *v.* （使）終止；（使）結束

They can't terminate the agreement until next year.

他們要到明年才能終止合約。

12. express or implied 明示或暗示

The company makes no warranties, express or implied, with respect to the ovens.

關於那些烤箱該公司沒有明示或暗示性的擔保。

Unit Three
Employment Contracts
聘雇合約

聘雇合約清楚載明雇主和員工對彼此之間的權利義務，讓雙方能有所遵循，避免因認知不同而引發紛爭。聘雇合約書可因工作性質的不同而有所差異。本合約書是針對外國企業聘雇台籍人士到海外工作，雙方所簽署的合約。內容主要涵蓋職務說明、任期、工時、薪資及福利、競業禁止、職場紀律等條款。

✔ Learning Goals

1. 了解聘雇合約書的格式、寫法及主要的內容

2. 針對海外工作的聘雇合約書要特別說明食、住、行、醫療、保險等生活上的照顧

Sample 範本

BRIDGES COLD ROLLING MILLS

15th FLOOR, US Steel Tower
600 Grant Street
Pittsburgh, PA, 15222
USA
TELEPHONE: 412-641-6500
FACSIMILE: 412-641-6956
E-MAIL: personnel@bridgescold.com

中譯

15222 美國賓州匹茲堡市
格蘭特街 600 號
美國鋼鐵大廈 15 樓
電話：412-641-6500
傳真：412-641-6956
電子郵件：personnel@
bridgescold.com

A G R E E M E N T
with
Group A Employee

A 級員工聘雇合約

This Agreement by and between BRIDGES COLD ROLLING MILLS of PITTSBURGH, PENNSYLVANIA (hereafter "the Company") and Morris Wang (hereafter "the Employee") is dated the 15th day of September, 20XX.

本合約為賓州匹茲堡市的 BRIDGES COLD ROLLING MILLS （以下稱「本公司」）和王莫里斯（以下稱「該職員」）於二零 XX 年九月十五號所簽訂的。

ARTICLE 1 — EMPLOYMENT

第一條 聘雇

1.1 <u>Employment</u>

Subject to the terms and conditions set forth herein, the Company **hereby**[1] employs the Employee, who hereby accepts employment with the Company.

1.1 <u>聘雇</u>

依此載明之條款與條件，本公司茲聘雇該員工，該員工也同意受雇於本公司。

1.2 <u>Position and Duties</u>

The Employee shall serve in the position of Turn Foreman for the Company. The Employee upon arrival at the Company's Williamsport, Pennsylvania plant shall **devote**[2] all his productive time and attention to the Company's operations, and shall use his best efforts in the **discharge**[3] of his duties as required by the position and the Company's regulations.

1.2 <u>職稱與職責</u>

該員工於本公司擔任值班領班一職。向本公司位在賓州威廉斯波特的工廠報到後，該員工應全職投入公司的營運，並盡全力履行其職位與本公司規定所要求的職責。

299

1.3 Term

The term of employment under this Agreement shall be 36 months, from April 1, 20XX to March 31, 20XX, unless otherwise extended by mutual **consent**.[4] It shall include training in Pittsburgh and then duties in Williamsport **inclusive of**[5] a three-month probationary period upon arrival.

ARTICLE 2 — COMPENSATION

2.1 Working Hours and Rest Days

Normal working days for the Employee shall be all the days of the week, except Saturdays, Sundays, and public holidays. Additional days off will be provided at the discretion of the Company. The normal working hours of each working day shall follow the Company attendance regulation, but not exceed eight working hours.

2.2 Salary

The Company shall pay the Employee a starting salary (basic pay plus allowance) of US$4,200 per month. The salary is to be adjusted and / or increased in accordance with the conditions set forth in the employee payment **provisions**[6] of the Company decree.

2.3 Overseas Allowance

The Company shall pay the Employee an additional overseas allowance of US$1,400 per month upon arrival in Pittsburgh.

2.4 Relocation Allowance

Should the Employee be accompanied by his family, the Company shall pay him a relocation allowance of US$3,800. The Company shall pay the Employee half that amount if he is not accompanied by his family.

1.3 期限

除非經雙方同意予以延長，本合約的受雇期為三十六個月，從二零 XX 年四月一號至二零 XX 年三月三十一號。該期限應包含在匹茲堡的受訓和之後在威廉斯波特的正式工作，報到後三個月的試用期亦包括在內。

第二條 薪水與福利

2.1 工作時數與休息日

除了星期六、日和國定假日之外，每天皆為該員工的正常工作日。本公司將斟酌是否提供額外休假。每個工作日之正常工時應遵照本公司的出席規定，唯不得超過八小時。

2.2 薪資

本公司應支付該員工每個月四千兩百元美金的起薪（底薪加津貼）。薪水將依公司規範中的員工薪資條款所提出之情況予以調整和（或）增加。

2.3 海外津貼

該員工到匹茲堡報到後，本公司應額外支付每個月一千四百元美金的海外津貼。

2.4 調遷津貼

若該員工有家眷陪同，本公司應支付他三千八百元美金的調遷津貼；若無家眷同行，本公司應支付上述金額的一半。

2.5 Overtime Allowance

A. Overtime for the Employee shall be recovered by leave on an hourly basis at a schedule mutually acceptable to the Employee and his immediate supervisor. If a mutually acceptable schedule cannot be agreed upon, such overtime shall be added to his home-leave period, with a maximum of seven days per 12 months.

B. Overtime hours shall be counted as those worked **in excess of**[7] the normal working hours as defined in Article 2.1.

2.6 Living Allowance

The Company shall pay the Employee a living allowance of US$15 per calendar day on-site and for business travel.

2.7 Meal Allowance

A. The Company shall pay a meal allowance of US$15 per calendar day on-site and for business travel to the Employee. Such an allowance shall not be paid for any day he is not working due to personal needs (i.e. personal absence), sick leave exceeding six days per year, and home leave.

B. If the Company arranges for the Employee to lodge in a hotel, the meal allowance shall be adjusted to cover his actual expenditure for meals.

2.8 Accommodation

Should the Employee be accompanied by his family, the Company shall provide him with a house. The Company shall provide the Employee not accompanied by his family with a private room in a house shared with another employee. All housing shall be **furnished**.[8] The Employee shall, however, be expected to pay for the **utilities**.[9]

2.5 加班津貼

A. 該員工加班可用以每小時為單位的補休來補償,補休時間需經該員工及其直屬主管雙方同意。若雙方對補休時間未能達成共識,該加班時數可併入探親假,以每十二個月最多七天為上限。

B. 超過條文 2.1 所定義的正常工時才算加班。

2.6 生活津貼

本公司應於該員工駐廠工作及出差時支付其每日十五元美金之生活津貼。

2.7 伙食津貼

A. 本公司應於該員工駐廠工作和出差時支付其每日十五元美金之伙食津貼。因個人需求(如事假)、每年超過六天以上的病假及探親假而未上班的天數,將不支付該津貼。

B. 若本公司安排該員工入住飯店,伙食津貼得予以調整以支付其實際的餐費。

2.8 住所

若該員工有家眷陪同,本公司應提供其一間房子;若該員工無家眷陪同,本公司應提供其與另一名員工共住但擁有個人房間的房子。所有的住所皆需配置傢俱,但該員工需支付瓦斯及水電的費用。

2.9 Transportation

The Employee shall be provided with a vehicle for the duration of his contract.

2.10 Insurance

The Company shall provide any insurance coverage for the Employee required under federal or state law.

2.11 Leave

The Employee is entitled to the following:

A. If not accompanied by his family, one home leave of two weeks after five and a half months' absence from Taiwan.

B. If accompanied by his family, one home leave of four weeks after 11 months' absence from Taiwan.

C. Travel on leave by economy class between Pittsburgh and Taoyuan for himself and his family if they are accompanying him.

D. Six days of sick leave with pay per year. If personal leave or sick leave is in excess of six days, his monthly salary and overseas allowance shall be subject to a **pro rata**[10] deduction based on a 30-day month.

The leave schedule shall be arranged as far as possible in advance so as to avoid a **hindrance**[11] to plant operations.

2.12 Medical Allowance

The Company shall pay or reimburse medical and hospital expenses **incurred**[12] in the United States and / or while traveling on business to the Employee and his family (wife and dependent children only).

2.9 交通

合約期間該員工應予以配車。

2.10 保險

本公司應提供該員工任何聯邦法或州法所要求之保險項目。

2.11 休假

該員工可享有下列：

A. 若無家眷陪伴，於離台五個半月後有兩週的探親假。

B. 若家眷同行，於離台十一個月後有四週的探親假。

C. 若該員工有家眷陪同，休假時其與家人可搭經濟艙往返匹茲堡和桃園。

D. 每年六天給薪病假。若事假或病假超過六天，以每月三十日計算，按比例扣減其月薪與海外津貼。

請假應儘早事先安排，以免妨礙工廠營運。

2.12 醫療津貼

本公司應支付或補償該員工及其家人（限配偶和未成年子女）發生在美國和（或）出差期間的看病和住院開銷。

2.13 Business Travel Allowance

The Company shall provide business-class tickets to the Employee for business travel and to the Employee, his spouse, and children for travel of reporting for and leaving duty between Pittsburgh and Taoyuan. The Company shall pay or reimburse reasonable travel expenses including hotels and meals not covered by the meal allowance set forth in Article 2.7.

2.13 出差津貼

本公司應於該員工出差時提供商務艙機票；該員工報到及離職時，本公司亦需提供其與配偶、子女往返匹茲堡和桃園的商務艙機票。本公司應支付或補償未涵蓋在條文 2.7 所載明之伙食津貼內的合理旅費，包含飯店和用餐的費用。

ARTICLE 3 — TERMINATION

第三條 解約

3.1 Death / **Disability**[13]

The Company may elect to terminate this Agreement and the Employee's employment hereunder **forthwith**[14] upon the occurrence of any of the following events:

A. The death or legal incapacity of the Employee.

B. Any mental or physical disability of the Employee that prevents him from discharging his duties as outlined in this Agreement.

C. The **breach**[15] of this Agreement by the Employee.

D. The dishonesty, **willful**[16] breach of duty, or **habitual**[17] neglect of duty by the Employee in the course of his employment.

3.1 死亡 / 無行為能力

一旦發生下列任一事件，本公司得依此選擇立即終止這份合約及該員工之聘雇：

A. 該員工死亡或喪失法律上的行為能力。

B. 任何身心障礙導致該員工無法履行本合約所概述的職責。

C. 該員工違反本合約。

D. 該員工於任職期間不誠實、故意違反職責或習慣性的怠忽職守。

3.2 Early Termination

The Company and / or the Employee may terminate this employment before the expiry date by giving at least three months' notice. In case early termination is initiated by the Company for any reason other than **pursuant to**[18] Article 3.1, the Company shall pay three months' salary to the Employee.

3.2 提前解約

在合約到期日前欲終止此聘雇關係，本公司和（或）該員工需三個月前告知。若本公司非依據條文 3.1 之理由而提出提前解約，應支付該員工三個月的薪資。

ARTICLE 4 — MISCELLANEOUS

4.1 Nondisclosure

The Employee shall not, either before or after termination of his employment with the Company, disclose to a third party any confidential information relating to the Company. The Employee recognizes and acknowledges that any financial information concerning the Company's customers is confidential.

All work-related property, including, without limitation, books, manuals, records, reports, notes, contracts, lists, and blueprints (or copies **thereof**),[19] and equipment furnished to or prepared by the Employee during his employment, belongs to the Company and shall be promptly returned to the Company upon termination of employment.

Following termination, the Employee shall not retain any written or other **tangible**[20] material containing proprietary information of the Company. **Notwithstanding**[21] any other provision hereof, Article 4.1 shall survive termination of this Agreement and the Employee's employment hereunder and continue in effect for a period of three years thereafter.

4.2 Noncompetition

During the term of his employment under this Agreement, the Employee shall not, directly or indirectly, promote, engage, or participate in any existing or **prospective**[22] business in competition with the Company, its subsidiaries, or affiliates.

4.3 Compliance

The Employee shall observe all rules and regulations of the Company **applicable**[23] to its officers and shall conduct himself at all times with due regard to public **conventions**[24] and **morals**.[25] The Employee shall not commit any act that would harm the Company or industry / industries in which the Company is involved.

4.4 Governing Law

第四條 雜項

4.1 保密

該員工在終止與本公司的聘雇關係之前或之後,不應向第三者公開有關本公司的任何機密資訊。該員工承認並同意任何有關本公司客戶的財務資訊皆屬機密。

所有與工作相關的所有物包括但不限於書籍、手冊、記錄、報告、筆記、合約、清單和藍圖(或其副本),以及任職期間提供給該員工或由其所準備之器材皆屬於本公司的財產,需於聘雇關係終止時立即歸還給本公司。

聘雇關係終止後,該員工不應保留任何內含本公司私有資訊之書面或其他實體的資料。儘管有其他條款的規定,條文4.1 在本合約與聘雇關係終止之後的三年仍具有效力。

4.2 競業禁止

在本合同所約定之聘雇期間,該員工不得直接或間接促進、從事或參與任何與本公司、分公司或子公司競爭之現存或未來事業。

4.3 遵照規定

該員工應遵從所有適用於本公司職員之規定與規章,並隨時約束自己的行為舉止使其符合社會常規與道德規範。該員工不應從事任何會損害公司或公司相關產業之行為。

This Agreement shall be governed by and **construed**[26] in accordance with the laws of the United States of America.

4.4 管轄法律

本合約應受美國法律管轄並依其作解釋。

4.5 <u>Taxes</u>

The Employee shall be responsible for paying his taxes in accordance with the law.

4.5 <u>稅賦</u>

員工應依法自行繳納稅金。

4.6 <u>Visas and Customs Clearance</u>

All expenses for the Employee and his family concerning visas and customs clearance will be on the Company, except import duties for personal belongings.

4.6 <u>簽證和通關</u>

除了私人物品的進口稅之外，員工及其家人所有關於簽證和通關之開銷皆由本公司負擔。

4.7 <u>**Arbitration**</u>[27]

Any breach of this Agreement or **controversy**[28] or claim relating to this Agreement shall be settled by arbitration in accordance with the commercial arbitration rules of the United States of America, and judgment on the **award**[29] **rendered**[30] by the arbitrator(s) may be entered in any court having **jurisdiction**.[31] Each party shall pay the expenses of its witnesses and all other expenses connected with presenting its case.

4.7 仲裁

任何違約或與本合約相關之爭議或要求，應依照美國商務仲裁規定提付仲裁，而對於仲裁人所作出的判決都可進入擁有司法管轄權的法院來審理。雙方應各自支付其證人和所有其他與出庭陳述案件相關之開銷。

In witness whereof,[32] **the parties hereto**[33] have executed this Agreement as of the day and the year first above written.

雙方自上述之日期起履行此合約，以此為證。

BRIDGES COLD ROLLING MILLS
Duncan Watson
Duncan Watson
General Manager

Employee
Morris Wang
Morris Wang

BRIDGES COLD ROLLING MILLS
鄧肯・華森
總經理

員工
王莫里斯

Vocabulary and Phrases

1. **hereby** [ˌhɪrˈbaɪ] *adv.* 特此；茲

2. **devote** [dɪˈvot] *v.* 將（時間、努力等）完全獻給

3. **discharge** [dɪsˈtʃɑrdʒ] *n.* 執行；履行（亦可當 *v.*）

4. **consent** [kənˈsɛnt] *n.* 同意；贊成

5. **inclusive of** 包括；包含

6. **provision** [prəˈvɪʒən] *n.* 規定；條款

7. **in excess of** 超過

8. **furnish** [ˈfɜnɪʃ] *v.* 為（房間）配置（傢俱等）；供應

9. **utility** [juˈtɪlətɪ] *n.* 公共設施
（此指「瓦斯及水電」；作此義解時，常用複數）

10. **pro rata** [pro ˈrɑtə] *adj.* 按比例的

11. **hindrance** [ˈhɪndrəns] *n.* 妨礙；障礙

12. **incur** [ɪnˈkɜ] *v.* 招致；帶來

13. **disability** [ˌdɪsəˈbɪlətɪ] *n.* 無能；殘障

14. **forthwith** [ˌforθˈwɪθ] *adv.* 立即；毫不拖延地

15. **breach** [britʃ] *n.* 違反；不履行

16. **willful** [ˈwɪlfəl] *adj.* 故意的；任性的（= wilful）

17. **habitual** [həˈbɪtʃuəl] *adj.* 習慣的；習以為常的

18. **pursuant to** 根據；依照

19. **thereof** [ðɛrˈəv] *adv.* 其；關於那

20. **tangible** [ˈtændʒəbl̩] *adj.* 實體的；有形的

21. **notwithstanding** [ˌnɑtwɪθˈstændɪŋ] *prep.* 儘管

22. **prospective** [prəˈspɛktɪv] *adj.* 未來的；即將發生的

23. **applicable** [ˈæplɪkəbl̩] *adj.* 適用的；可實施的

24. **convention** [kənˈvɛnʃən] *n.* 習俗；常規

25. **morals** [ˈmɔrəlz] *n.* 道德；倫理

26. **construe** [kənˈstru] *v.* 解釋；詮釋

27. **arbitration** [ˌɑrbəˈtreʃən] *n.* 仲裁；公斷

28. **controversy** [ˈkɑntrəˌvɜsɪ] *n.* 爭議

29. **award** [əˈwɔrd] *n.* 判決；裁定

30. **render** [ˈrɛndə] *v.* 提出；交付

31. **jurisdiction** [ˌdʒurɪsˈdɪkʃən] *n.* 審判權；管轄權

32. **in witness whereof** 以此為證；特立此證
（= in testimony whereof）

33. **the parties hereto** （此指）本合同雙方

Notes

Notes

Notes

Notes

Notes

Notes

Notes

Notes

Notes

Practical Model Business Correspondence

Clear · Concise · Complete

商務實戰 英語書信實例

發行人	鄭俊琪
總編輯	陳豫弘
副總編輯	吳嘉玲
作者	莊錫宗 · LiveABC
責任編輯	廖珮筠
英文編輯	Patrick Cowsill · Mike Corsini
美術設計	張佳琦
出版發行	希伯崙股份有限公司
	105 北市松山區八德路 3 段 32 號 12 樓
	劃撥：1939-5400
	電話：(02) 2578-7838
	傳真：(02) 2578-5800
	電子郵件：Service@LiveABC.com
出版日期	民國 99 年 10 月 初版五刷
法律顧問	朋博法律事務所
印刷	禹利電子分色有限公司

本書若有缺頁、破損、裝訂錯誤，請寄回本公司更換

版權所有，禁止以任何方式，在任何地方作全部或局部翻印、仿製或轉載

Copyright © 2009 by LiveABC Interactive Corporation
All rights reserved. Printed in Taiwan

《商務實戰 英語書信實例》讀者回函卡

謝謝您購買 LiveABC 互動英語系列產品

如果您願意，請您詳細填寫下列資料，免貼郵票寄回 LiveABC 即可獲贈《CNN 互動英語》、《Live 互動英語》、《每日一句週報》電子學習報 3 個月期（價值：900 元）及 LiveABC 不定期提供的最新出版資訊。

姓名		性別 □男 □女

出生日期	年 月 日

住址	□□□ 聯絡電話

E-mail	

學歷	□國中以下 □國中 □高中 □大專及大學 □研究所

職業	□學生 □資訊業 □工 □商 □服務業 □軍警公教 □自由業及專業 □其他 _____

您從何處得知本書？
□書店 □網站
□電子型錄 □他人推薦
□雜誌
□其他 _____

您以何種方式購得此書？
□一般書店 □連鎖書店
□網路 □郵局劃撥
□其他

您覺得本書的價格？
□偏低 □合理 □偏高

您對本書的評價

	書名	封面	內容	編排	紙張
很滿意	□	□	□	□	□
還不錯	□	□	□	□	□
普通	□	□	□	□	□
不滿意	□	□	□	□	□
很後悔	□	□	□	□	□

您希望我們製作哪些學習主題？

您對我們的建議：

書 貼 處

廣 告 回 信
台 北 郵 局 登 記 證
台 北 廣 字 1194 號

縣
市

市
區
鄉
鎮

村
里
路
街

段

鄰
巷

弄

號

樓

室

希伯崙股份有限公司客戶服務部 收

1 0 5

台北市松山區八德路三段32號12樓

LiveABC

英語數位學習第一品牌